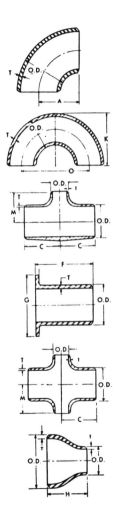

GUIDE TO FLANGES, FITTINGS, AND PIPING DATA
Second Edition

R. R. LEE

Published by

Gulf Publishing Company
Houston, London, Paris, Zurich, Tokyo

in association with

Plant Engineering

POCKET GUIDE TO FLANGES, FITTINGS, AND PIPING DATA
Second Edition

Copyright © 1984, 1992 by Gulf Publishing Company, Houston, Texas. All rights reserved. Printed in the United States of America. This book, or parts thereof, may not be reproduced in any form without permission of the publisher.

Gulf Publishing Company
Book Division
P.O. Box 2608, Houston, Texas 77252-2608

10 9 8 7 6 5 4 3 2

Library of Congress Cataloging-in-Publication Data

Lee, R. R. (Robert R.), 1928–
    Pocket guide to flanges, fittings, and piping data/R. R. Lee.—2nd ed.
        p.   cm.
    Includes index.
    ISBN 0-88415-023-2
    1. Flanges—Handbooks, manuals, etc. 2. Pipe-fittings— Handbooks, manuals, etc. 3. Pipe—Handbooks, manuals, etc. I. Title.
    TA492.F5L44   1992
    621.8′672—dc20   91-36552
                                                       CIP

Printed on Acid-Free Paper (∞)

To my daughter E. Doris Lee

CONTENTS

Appendix A

Appendix B

ACKNOWLEDGMENTS

I would like to express my appreciation to certain colleagues and business associates for their contributions and support for strong material control programs: J. M. Smith, P. Rajagapolan, Mohd Kunhi, MIS Dubai, U.A.E.; A. S. Zeidy, Cairo, Egypt; T. W. Acosta, C. L. Davis, McDermott (Saudi Arabia); Ralph Williams, Citizens National Bank of Texas; Edgar H. Von Minden, Jr., Tube Turns Technologies, Inc.; Beatrice Welch, M. D.; J. R. Lee, AMOCO Production Co.; C. H. Haslam, OEA (Oman); Delmar & Elmar Boyd, Bechtel (Libya); Kevin Talbot, HAH (Kuwait); Yousef A. Al-Omani, MISCO-Kuwait; and to my wife Pat for her assistance with the preparation of this book.

▌PREFACE

Materials personnel are usually the first to be assigned to a project and the last to close it. Their responsibilities of collecting tables, catalogs, specifications, and materials accounting system forms and supplies commence even before they depart for the jobsite, which could very well be in Mukluk, Al Mukalla, Al Khobar, Belo Horizonte, or just outside of Houston. To have all the required documentation on hand at the jobsite is a real challenge.

This handbook is designed to bridge the gap for piping data and materials storage procedures until a more formal specification has been issued by your project manager. It is intended to help train new materials personnel on the project by answering questions they may be afraid to ask. The markings on fittings and pipe are explained in a non-technical language. Pipe schedules are provided to show equal schedules in certain sizes, but different call-outs such as standard and schedule 40. Tables describing the size and length of bolts for flanges and clamps are included, in addition to the size wrench required to fit the nut. An oval ring gasket cross-reference chart is also included.

The book is quite useful to drafters, purchasing agents, pipefitters, students, and project managers.

The material in the book is believed to be technically correct; neither the author nor Gulf Publishing Company warrant its use. Always consult the applicable specification issued by the project manager at your project.

R. R. Lee
Houston, Texas

1

ANSI FLANGES

A flange is used to join pipe, valves, or a vessel within a system. The common ANSI flanges are shown in Figure 1-1, and special flanges are illustrated and defined later in this chapter.

ANSI Standards

Pressure ratings for flanges are designed to ANSI standards of 150 lb, 300 lb, 400 lb, 600 lb, 900 lb, 1500 lb, and 2500 lb. The most common terminology used is the pound reference, although the more formal reference is by class, such as Class 150 flange.

The ANSI standards require that each flange be stamped with identifying markings as shown in Figure 1-2. The markings include:

1. Manufacturer's trade name.
2. Nominal pipe size—the outside diameter of pipe the flange will match when welded to the pipe.

Welding Neck Flange

Slip-on Flange

Threaded Flange

Lap Joint Flange

(Continued on next page)

Reducing Slip-on Flange

Socket Welding Flange

Blind Flange

Figure 1-1. ANSI flanges. (Courtesy of Taylor Forge.)

3

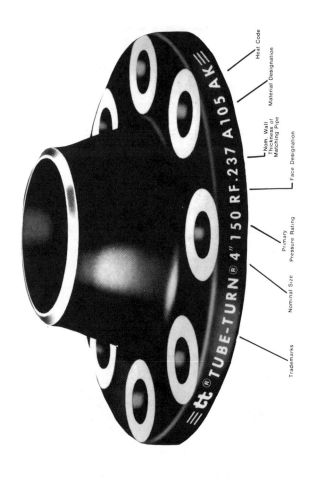

Figure 1-2. Flange identification markings. (Courtesy of Tube Turns Technologies, Inc.)

4

3. Primary pressure rating (also known as the flange rating)—150-lb, 300-lb, etc.
4. Face designation—the machined gasket surface area of the flange (resembles a phonograph record, see Figure 1-3). The flange face is the most important part of the flange. The $1/16$-in. raised face is

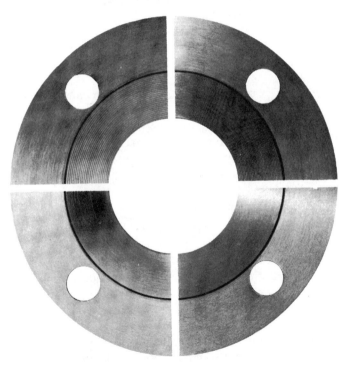

Figure 1-3. Flange face gasket surfaces. (Courtesy of Taylor Forge.)

common in 150-lb and 300-lb classes. Heavier ratings are ¼-in. raised faces. A ring type joint is available in all classes, but more common in the 600-lb and greater classes.

5. Bore (also known as the nominal wall thickness of matching pipe)—the measure of the flange wall thickness, which matches the inside dimension of the pipe being used.

6. Material designation—ASTM specifications that describe the raw materials from which the flange is made, such as ingots, blooms, billets, slabs, or bars.

7. Ring gasket number—used when the flange face is a ring type joint style.

8. Heat number or code—the batch number used by steel forgers to identify a particular batch number of steel forgings and test results. The mill test results are made available to the purchasers of the flanges.

Flange Bores

Weldneck and socket weld flanges are drilled (machined) with the wall thickness of the flange having the same dimensions of the matching pipe. The lighter the pipe is, the larger the bore; conversely, the heavier the pipe, the smaller the bore.

Other flanges are drilled to match the outside diameter of pipe sizes, and do not have bore markings to indicate a pipe schedule.

Types of ANSI Flanges

Weldneck Flange

This flange, shown in Figures 1-1 and 1-2, is normally used for high-pressure, cold or hot temperatures.

Slip-on and Lap Joint Flanges

Figure 1-4 shows these "twin" flanges. Note, however, that a slip-on flange is bored slightly larger than the OD of the matching pipe. The pipe slips into the flange prior to welding both inside and outside to prevent leaks.

The lap joint flange has a curved radius at the bore and face to accommodate a lap joint stub end. (Stub ends are described in Chapter 2.) The lap joint flange and stub end assembly are normally used in systems requiring frequent dismantling for inspection.

Threaded Flange

This type of flange, shown in Figure 1-5, is used in systems not involving temperature or stresses of any magnitude.

Socket Weld Flange

This flange is similar to slip-on flange, except they have a bore and a counter bore. See Figure 1-6.

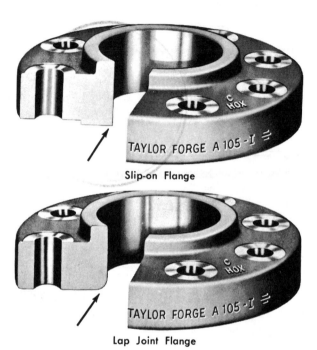

Slip-on Flange

Lap Joint Flange

Figure 1-4. Slip-on and lap joint flanges. (Courtesy of Taylor Forge.)

The counter bore is slightly larger than the OD of the matching pipe, allowing the pipe to be inserted. A restriction is built into the bottom of the bore, which acts as a shoulder for the pipe to rest on, and has the same ID of the matching pipe. The flow is not restricted in any direction.

8

Figure 1-5. Threaded flange. (Courtesy of Hackney, Inc., a division of Trinity Industries.)

Figure 1-6. Socket weld flange. (Courtesy of Hackney, Inc., a division of Trinity Industries.)

Reducing Flange

The reducing flange is similar in every respect to the full size of the flange from which the reduction is to be made. See Figure 1-7A.

Figure 1-7. Reducing and blind flanges. (Courtesy of Tube Turns Technologies, Inc.)

The reducing flange is described in the same manner as a reducer—that is, the large end first, the reduction second. An example would be a 6-in. raised face by 2-in. threaded reducing flange, ANSI 150 lb rating.

Blind Flange

Figure 1-7B shows a blind flange, which has no bore, and is used to close ends of piping systems. A blind flange also permits easy access to a line once it has been sealed.

The blind flange is sometimes machined to accept a pipe of the nominal size to which the reduction is being made. The reduction can be either threaded or welded.

Miscellaneous Flanges

Long Weldneck Flange

This is a special flange used for nozzles on pressure vessels. The hub is always straight, and the hub thickness is greater than the diameter of any piping that may be bolted to the flange. See Figure 1-8.

Figure 1-8. Long weldneck flange. (Courtesy of Hackney, Inc., a division of Trinity Industries.)

Orifice Flange

The function of an orifice flange is to meter the flow of liquids and gases through a pipe line. Figure 1-9 shows an orifice flange complete with bolting and jack screws. The jack screws are used to spread the flanges apart in a line to change an orifice plate between the two flanges.

Figure 1-9. Orifice flange with jack screws. (Courtesy of Hackney, Inc., a division of Trinity Industries.)

Figure 1-10. Cutaway of an orifice flange. (Courtesy of Taylor Forge.)

As illustrated in Figure 1-10, the orifice flange has drilled and tapped outlets for installing the metering device (recorder).

Materials Control

Receiving Flanges

It is very important to carefully examine every flange to verify that each conforms to the purchase order specifications. It is equally important to check for damage to the face and beveled end, and inspect the flanges for corrosion. Report any damage or other discrepancies to your supervisor or prepare an over, short, and damage report (OS&D).

Applying Commodity Code Numbers

When flanges are received, apply your company's commodity identification numbers on flanges and other materials.

Apply the numbers in the same area as the manufacturer's stamped flange identification numbers. Always stock the flanges with the numbers facing the same direction for prompt and positive material identification.

Storing Flanges

The preferred storage method for flanges is face down on wooden pallets or wooden docks, using dunnage to separate each layer of flanges, as shown in Figure 1-11. (For dunnage, use wooden strips, paneling, cardboard, plywood, etc.) Special care should be extended to the flange facing during storage or shipments to other areas.

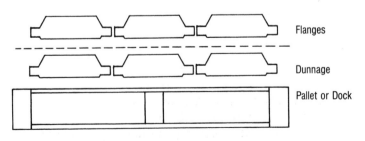

Flanges

Dunnage

Pallet or Dock

Figure 1-11. Diagram of method for storing flanges.

Accounting Records

Good technical descriptions are necessary in any system. The very best source for technical descriptions is your company's computer master. If the computer master is not available, then try to be consistent each time you describe the same materials in the accounting system. A good technical description of the weldneck flange shown in Figure 1-2 would be:

What is it?	Weldneck Flange.
What type facing?	Faced and drilled raised face.
What is the pressure rating?	ANSI 150 lb rating.
What is the bore?	Bore standard weight (.237-inch wall thickness).

It is not necessary when describing flanges and other materials to include the trademark, the ASTM specification, or the heat number unless requested to do so by your supervisor.

Shipping Flanges

When flanges are to be shipped to other locations either loose or as an integral part of a fabricated pipe piece, protect the flange face with a flange protector, and the beveled end with an end protector.

Flange protectors are very inexpensive compared to replacing even one damaged flange during an offshore

Figure 1-12. Commercial flange protectors. (Courtesy of Mark V, division of Phoenix Industries, Inc.)

or overseas shipment. Figure 1-12 shows a low-cost, durable protector that can be quickly installed and locked in place by friction lock poly bolts.

These protectors protect the entire flange facing from impact damage, and will not deteriorate. The protectors are reusable. A wrap of duct tape around the outer edge of the protector and the flange ring will prevent sand and water from entering the pipe or nozzle areas.

Flange Dimensions

Table 1-1 includes the wall thickness schedules and dimensions of ANSI flanges. The table is very useful to engineers, draftsmen, fabricators, purchasing, and materi-

Table 1-1
Dimensions for ANSI Flanges
(Courtesy of Taylor Forge)

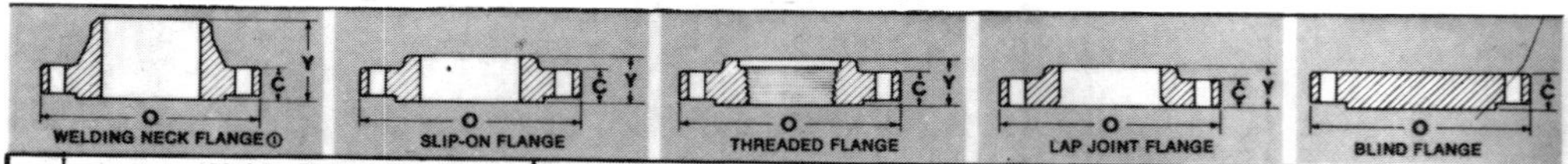
WELDING NECK FLANGE ①
SLIP-ON FLANGE
THREADED FLANGE
LAP JOINT FLANGE
BLIND FLANGE
O
C
Y

150 LB. FLANGES

Nom. Pipe Size	O	C ②	Y ② Weld Neck	Y ② Slip on Thrd.	Y ② Lap Joint	Bolt Circle	No. and Size of Holes
1/2	3 1/2	7/16	1 7/8	5/8	5/8	2 3/8	4-5/8
3/4	3 7/8	1/2	2 1/16	5/8	5/8	2 3/4	4-5/8
1	4 1/4	9/16	2 3/16	11/16	11/16	3 1/8	4-5/8
1 1/4	4 5/8	5/8	2 1/4	13/16	13/16	3 1/2	4-5/8
1 1/2	5	11/16	2 7/16	7/8	7/8	3 7/8	4-5/8
2	6	3/4	2 1/2	1	1	4 3/4	4-3/4
2 1/2	7	7/8	2 3/4	1 1/8	1 1/8	5 1/2	4-3/4
3	7 1/2	15/16	2 3/4	1 3/16	1 3/16	6	4-3/4
3 1/2	8 1/2	15/16	2 13/16	1 1/4	1 1/4	7	8-3/4
4	9	15/16	3	1 5/16	1 5/16	7 1/2	8-3/4
5	10	15/16	3 1/2	1 7/16	1 7/16	8 1/2	8-7/8
6	11	1	3 1/2	1 9/16	1 9/16	9 1/2	8-7/8
8	13 1/2	1 1/8	4	1 3/4	1 3/4	11 3/4	8-7/8
10	16	1 3/16	4	1 15/16	1 15/16	14 1/4	12-1
12	19	1 1/4	4 1/2	2 3/16	2 3/16	17	12-1
14	21	1 3/8	5	2 1/4	3 1/8	18 3/4	12-1 1/8
16	23 1/2	1 7/16	5	2 1/2	3 7/16	21 1/4	16-1 1/8
18	25	1 9/16	5 1/2	2 11/16	3 13/16	22 3/4	16-1 1/4
20	27 1/2	1 11/16	5 11/16	2 7/8	4 1/16	25	20-1 1/4
24	32	1 7/8	6	3 1/4	4 3/8	29 1/2	20-1 3/8

300 LB. FLANGES

Nom. Pipe Size	O	C ②	Y ② Weld Neck	Y ② Slip on Thrd.	Y ② Lap Joint	Bolt Circle	No. and Size of Holes
1/2	3 3/4	9/16	2 1/16	7/8	7/8	2 5/8	4-5/8
3/4	4 5/8	5/8	2 1/4	1	1	3 1/4	4-3/4
1	4 7/8	11/16	2 7/16	1 1/16	1 1/16	3 1/2	4-3/4
1 1/4	5 1/4	3/4	2 9/16	1 1/16	1 1/16	3 7/8	4-3/4
1 1/2	6 1/8	13/16	2 11/16	1 3/16	1 3/16	4 1/2	4-7/8
2	6 1/2	7/8	2 3/4	1 5/16	1 5/16	5	8-3/4
2 1/2	7 1/2	1	3	1 1/2	1 1/2	5 7/8	8-7/8
3	8 1/4	1 1/8	3 1/8	1 11/16	1 11/16	6 5/8	8-7/8
3 1/2	9	1 3/16	3 3/16	1 3/4	1 3/4	7 1/4	8-7/8
4	10	1 1/4	3 3/8	1 7/8	1 7/8	7 7/8	8-7/8
5	11	1 3/8	3 7/8	2	2	9 1/4	8-7/8
6	12 1/2	1 7/16	3 7/8	2 1/16	2 1/16	10 5/8	12-7/8
8	15	1 5/8	4 3/8	2 7/16	2 7/16	13	12-1
10	17 1/2	1 7/8	4 5/8	2 5/8	3 3/4	15 1/4	16-1 1/8
12	20 1/2	2	5 1/8	2 7/8	4	17 3/4	16-1 1/4
14	23	2 1/8	5 5/8	3	4 3/8	20 1/4	20-1 1/4
16	25 1/2	2 1/4	5 3/4	3 1/4	4 3/4	22 1/2	20-1 3/8
18	28	2 3/8	6 1/4	3 1/2	5 1/8	24 3/4	24-1 3/8
20	30 1/2	2 1/2	6 3/8	3 3/4	5 1/2	27	24-1 3/8
24	36	2 3/4	6 5/8	4 3/16	6	32	24-1 5/8

400 LB. FLANGES

Nom. Pipe Size	O	C ②	Y ② Weld Neck	Y ② Slip on Thrd.	Y ② Lap Joint	Bolt Circle	No. and Size of Holes
1/2	3 3/4	9/16	2 1/16	7/8	7/8	2 5/8	4-5/8
3/4	4 5/8	5/8	2 1/4	1	1	3 1/4	4-3/4
1	4 7/8	11/16	2 7/16	1 1/16	1 1/16	3 1/2	4-3/4
1 1/4	5 1/4	13/16	2 5/8	1 1/8	1 1/8	3 7/8	4-3/4
1 1/2	6 1/8	7/8	2 3/4	1 1/4	1 1/4	4 1/2	4-7/8
2	6 1/2	1	2 7/8	1 7/16	1 7/16	5	8-3/4
2 1/2	7 1/2	1 1/8	3 1/8	1 5/8	1 5/8	5 7/8	8-7/8
3	8 1/4	1 1/4	3 1/4	1 13/16	1 13/16	6 5/8	8-7/8
3 1/2	9	1 3/8	3 3/8	1 15/16	1 15/16	7 1/4	8-1
4	10	1 3/8	3 1/2	2	2	7 7/8	8-1
5	11	1 1/2	4	2 1/8	2 1/8	9 1/4	8-1
6	12 1/2	1 5/8	4 1/16	2 1/4	2 1/4	10 5/8	12-1
8	15	1 7/8	4 5/8	2 11/16	2 11/16	13	12-1 1/8
10	17 1/2	2 1/8	4 7/8	2 7/8	4	15 1/4	16-1 1/4
12	20 1/2	2 1/4	5 3/8	3 1/8	4 1/4	17 3/4	16-1 3/8
14	23	2 3/8	5 7/8	3 5/16	4 5/8	20 1/4	20-1 3/8
16	25 1/2	2 1/2	6	3 11/16	5	22 1/2	20-1 1/2
18	28	2 5/8	6 1/2	3 7/8	5 3/8	24 3/4	24-1 1/2
20	30 1/2	2 3/4	6 5/8	4	5 3/4	27	24-1 5/8
24	36	3	6 7/8	4 1/2	6 1/4	32	24-1 7/8

600 LB. FLANGES

Nom. Pipe Size	O	C ②	Y ② Weld Neck	Y ② Slip on Thrd.	Y ② Lap Joint	Bolt Circle	No. and Size of Holes
1/2	3 3/4	9/16	2 1/16	7/8	7/8	2 5/8	4-5/8
3/4	4 5/8	5/8	2 1/4	1	1	3 1/4	4-3/4
1	4 7/8	11/16	2 7/16	1 1/16	1 1/16	3 1/2	4-3/4
1 1/4	5 1/4	13/16	2 5/8	1 1/8	1 1/8	3 7/8	4-3/4
1 1/2	6 1/8	7/8	2 3/4	1 1/4	1 1/4	4 1/2	4-7/8

900 LB. FLANGES

Nom. Pipe Size	O	C ②	Y ② Weld Neck	Y ② Slip on Thrd.	Y ② Lap Joint	Bolt Circle	No. and Size of Holes
1/2	4 3/4	7/8	2 3/8	1 1/4	1 1/4	3 1/4	4-7/8
3/4	5 1/8	1	2 3/4	1 3/8	1 3/8	3 1/2	4-7/8
1	5 7/8	1 1/8	2 7/8	1 5/8	1 5/8	4	4-1
1 1/4	6 1/4	1 1/8	2 7/8	1 5/8	1 5/8	4 3/8	4-1
1 1/2	7	1 1/4	3 1/4	1 3/4	1 3/4	4 7/8	4-1 1/8

1500 LB. FLANGES

Nom. Pipe Size	O	C ②	Y ② Weld Neck	Y ② Slip on Thrd.	Y ② Lap Joint	Bolt Circle	No. and Size of Holes
1/2	4 3/4	7/8	2 3/8	1 1/4	1 1/4	3 1/4	4-7/8
3/4	5 1/8	1	2 3/4	1 3/8	1 3/8	3 1/2	4-7/8
1	5 7/8	1 1/8	2 7/8	1 5/8	1 5/8	4	4-1
1 1/4	6 1/4	1 1/8	2 7/8	1 5/8	1 5/8	4 3/8	4-1
1 1/2	7	1 1/4	3 1/4	1 3/4	1 3/4	4 7/8	4-1 1/8

(Continued on next page)

Table 1-1 continued

	600 LB. FLANGES							900 LB. FLANGES							1500 LB. FLANGES							
2	6½	1	2⅞	1 7/16	1 7/16	5	8-¾	8½	1½	4	2¼	2¼	6½	8-1	8½	1½	4	2¼	2¼	6½	8-1	**2**
2½	7½	1⅛	3⅛	1⅝	1⅝	5⅞	8-⅞	9⅝	1⅝	4⅛	2½	2½	7½	8-1⅛	9⅝	1⅝	4⅛	2½	2½	7½	8-1⅛	**2½**
3	8¼	1¼	3¼	1 13/16	1 13/16	6⅝	8-⅞	9½	1½	4	2⅛	2⅛	7½	8-1	10½	1⅞	4⅝	2⅞	2⅞	8	8-1¼	**3**
3½	9	1⅜	3⅜	1 15/16	1 15/16	7¼	8-1	..	..	..	..	..	..	..	..	..	..	..	..	..	..	**3½**
4	10¾	1½	4	2⅛	2⅛	8½	8-1	11½	1¾	4½	2¾	2¾	9¼	8-1¼	12¼	2⅛	4⅞	3 9/16	3 9/16	9½	8-1⅜	**4**
5	13	1¾	4½	2⅜	2⅜	10½	8-1⅛	13¾	2	5	3⅛	3⅛	11	8-1⅜	14¾	2⅞	6⅛	4⅛	4⅛	11½	8-1⅝	**5**
6	14	1⅞	4⅝	2⅝	2⅝	11½	12-1⅛	15	2 3/16	5½	3⅜	3⅜	12½	12-1¼	15½	3¼	6¾	4 11/16	4 11/16	12½	12-1½	**6**
8	16½	2 3/16	5¼	3	3	13¾	12-1¼	18½	2½	6⅜	4	4½	15½	12-1½	19	3⅝	8⅜	5⅝	5⅝	15½	12-1¾	**8**
10	20	2½	6	3⅜	4⅜	17	16-1⅜	21½	2¾	7¼	4¼	5	18½	16-1½	23	4¼	10	6¼	7	19	12-2	**10**
12	22	2⅝	6⅛	3⅝	4⅝	19¼	20-1⅜	24	3⅛	7⅞	4⅝	5⅝	21	20-1½	26½	4⅞	11⅛	7⅛	8⅝	22½	16-2⅛	**12**
14	23¾	2¾	6½	3 11/16	5	20¾	20-1½	25¼	3⅜	8⅜	5⅛	6⅛	22	20-1⅝	29½	5¼	11¾	..	9½	25	16-2⅜	**14**
16	27	3	7	4 3/16	5½	23¾	20-1⅝	27¾	3½	8½	5¼	6½	24¼	20-1¾	32½	5¾	12¼	..	10¼	27¾	16-2⅝	**16**
18	29¼	3¼	7¼	4⅝	6	25¾	20-1¾	31	4	9	6	7½	27	20-2	36	6⅜	12⅞	..	10⅞	30½	16-2⅞	**18**
20	32	3½	7½	5	6½	28½	24-1¾	33¾	4¼	9¾	6¼	8¼	29½	20-2⅛	38¾	7	14	..	11½	32¾	16-3⅛	**20**
24	37	4	8	5½	7¼	33	24-2	41	5½	11½	8	10½	35½	20-2⅝	46	8	16	..	13	39	16-3⅝	**24**

	2500 LB. FLANGES						
½	5¼	1 3/16	2⅞	1 9/16	1 9/16	3½	4-⅞
¾	5½	1¼	3⅛	1 11/16	1 11/16	3¾	4-⅞
1	6¼	1⅜	3½	1⅞	1⅞	4¼	4-1
1¼	7¼	1½	3¾	2 1/16	2 1/16	5⅛	4-1⅛
1½	8	1¾	4⅜	2⅜	2⅜	5¾	4-1¼
2	9¼	2	5	2¾	2¾	6¾	8-1⅛
2½	10½	2¼	5⅝	3⅛	3⅛	7¾	8-1¼
3	12	2⅝	6⅝	3⅝	3⅝	9	8-1⅜
4	14	3	7½	4¼	4¼	10¾	8-1⅝
5	16½	3⅝	9	5⅛	5⅛	12¾	8-1⅞
6	19	4¼	10¾	6	6	14½	8-2⅛
8	21¾	5	12½	7	7	17¼	12-2⅛
10	26½	6½	16½	9	9	21¼	12-2⅝
12	30	7¼	18¼	10	10	24⅜	12-2⅞

WELDING NECK FLANGE BORES ①②

Nom. Pipe Size	Outside Diam.	Light Wall⑥	Sched. 20	Sched. 30	Std. Wall.	Sched. 40	Sched. 60	Extra Strong	Sched. 80	Sched. 100	Sched. 120	Sched. 140	Sched. 160	Double Extra Strong
½	0.840	.674	...	...	0.622	0.622	...	0.546	0.546	...	...	...	0.464	0.252
¾	1.050	.884	...	...	0.824	0.824	...	0.742	0.742	...	...	...	0.612	0.434
1	1.315	1.097	...	...	1.049	1.049	...	0.957	0.957	...	...	...	0.815	0.599
1¼	1.660	1.442	...	...	1.380	1.380	...	1.278	1.278	...	...	...	1.160	0.896
1½	1.900	1.682	...	...	1.610	1.610	...	1.500	1.500	...	...	...	1.338	1.100
2	2.375	2.157	...	...	2.067	2.067	...	1.939	1.939	...	...	...	1.687	1.503
2½	2.875	2.635	...	...	2.469	2.469	...	2.323	2.323	...	...	...	2.125	1.771
3	3.500	3.260	...	...	3.068	3.068	...	2.900	2.900	...	...	...	2.624	2.300
3½	4.000	3.760	...	...	3.548	3.548	...	3.364	3.364	...	...	...	...	2.728
4	4.500	4.260	...	...	4.026	4.026	...	3.826	3.826	...	3.624	...	3.438	3.152
5	5.563	5.295	...	...	5.047	5.047	...	4.813	4.813	...	4.563	...	4.313	4.063
6	6.625	6.357	...	...	6.065	6.065	...	5.761	5.761	...	5.501	...	5.187	4.897
8	8.625	8.329	8.125	8.071	7.981	7.981	7.813	7.625	7.625	7.437	7.187	7.001	6.813	6.875
10	10.750	10.420	10.250	10.136	10.020	10.020	9.750	9.750	9.562	9.312	9.062	8.750	8.500	8.750
12	12.750	12.390	12.250	12.090	12.000	11.938	11.626	11.750	11.374	11.062	10.750	10.500	10.126	10.750
14	14.000	13.500	13.376	13.250	13.250	13.124	12.812	13.000	12.500	12.124	11.814	11.500	11.188	...
16	16.000	15.500	15.376	15.250	15.250	15.000	14.688	15.000	14.312	13.938	13.564	13.124	12.812	...
18	18.000	17.500	17.376	17.124	17.250	16.876	16.500	17.000	16.124	15.688	15.250	14.876	14.438	...
20	20.000	19.500	19.250	19.000	19.250	18.812	18.376	19.000	17.938	17.438	17.000	16.500	16.062	...
24	24.000	23.500	23.250	22.876	23.250	22.624	22.062	23.000	21.562	20.938	20.376	19.876	19.312	...
30	30.000	29.376	29.000	28.750	29.250	...	...	29.000	...	...	...	...	...	...
36	36.000	35.376	35.000	34.750	35.250	34.500	...	35.000	...	...	...	...	...	...
42	42.000	...	...	...	41.250	...	...	41.000	...	...	...	...	...	...
48	48.000	...	...	...	47.250	...	...	47.000	...	...	...	...	...	...

NOTES:

① Always specify bore when ordering.

② Includes 1/16" raised face in 150 lb. and 300 lb. standards. Does **not** include 1/4" raised face in 400 lb. and heavier standards.

③ Inside pipe diameters are also provided by this table.

④ Other types, sizes and facings on application.

⑤ Stocked in carbon steel and a variety of other metals and alloys.

⑥ Light Wall diameters are identical to stainless steel Schedule 10S in sizes thru 12", and to Schedule 10 in sizes 14" and larger.

als control persons. Familiarize yourself with the table and its contents.

For an exercise, blank out the markings on a flange, and by using the table as a reference, properly identify the flange as to size, bolt holes, rating, bore, etc. The practice will enrich your skill and self-confidence as a materials person.

Table 1-1 also includes pipe schedules that describe the flange bores,as well as fittings. It is necessary at this time for you to become familiar with the following schedule terminology:

Light wall
Schedule 10 (Sch/10, S/10)
Schedule 20 (Sch/20, etc.)
Schedule 30
Schedule 40
Standard Weight
Schedule 60
Extra Strong (Extra heavy, EH, XH)
Schedule 80
Schedule 100
Schedule 120
Schedule 140
Schedule 160
Double Extra Strong (Double extra heavy, XXH, XXS)

Many of the schedules are identical in certain sizes, and either description is correct, but be consistent. An example of 6-in. Schedule 40, standard weight, or .237-

in. wall thickness. All three have the same meaning per Table 1-1, in the 6-in. size.

ANSI Flange Bolting

Stud Bolts

In ANSI piping, stud bolts are stamped with identifying numbers on the ends of bolts and the face of the nuts. A common mark for bolts is B7, and Grade 2H on the nuts. The bolts are often plated with various coatings, some of which are listed below:

Plating	Marking
Cadmium	B7
Zinc	L7
Chromium	B16
Tin	B8
Silver	B8M

Stud bolts are shipped from vendors as so many bolts/ nuts per pound. If the purchase order states 100 stud bolts with two heavy hex nuts each, then you may receive 99 or 102 bolts with nuts. To save time, and avoid counting each bolt, verify the total weight as being correct, verify about 25% of the markings per the specification, then store the bolts in the shipping container rather than using valuable warehouse shelving space.

Do not grease, spray, or clean bolts without specific instructions from your supervisor. Do not store new bolts with used bolts.

Machine Bolts

The machine bolt is commonly used for slip-on and threaded flanges. Only one heavy hex nut is required for a machine bolt. Both the bolt and the nut are identified the same way as for stud bolts.

Measuring Stud Bolts

A fast way to measure a stud bolt diameter is by measuring the thickness of the heavy hex nut; for example, a $1/2$-in. diameter bolt has a heavy hex nut that is $1/2$-in. thick.

Measure the length of stud bolts to the nearest $1/4$-in. from thread to thread, less the point heights as shown in Figure 1-13.

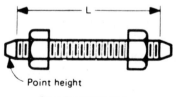

Stud Bolt With Nuts

Figure 1-13. Method of measuring stud bolts. (Courtesy of the American Petroleum Institute.)

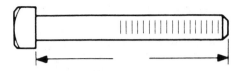

Figure 1-14. Method of measuring machine bolts. (Courtesy of the American Petroleum Institute.)

Measuring Machine Bolts

Measure the length of a machine bolt from the underside of the head to the end point. See Figure 1-14. (All bolts are rounded off to the nearest 1/4-in.).

Tables 1-2 through 1-16 describe stud bolt and machine bolt tables for all ANSI flanges previously discussed for raised and flat faced flanges, plus ring type joint flanges (RTJ). The tables also include the size of the wrench required to fit the heavy hex nuts used on each diameter of stud bolts. (Wrench size is calculated as one and one-half times the size of the bolt plus one eighth of an inch). Figure 1-15 shows a dimensional gauge for bolting.

Example: 1 1/2 × 1 = 1 1/2 + 1/8 = 1 5/8-in. wrench for a 1-in. nut.

Table 1-17 gives suggested materials for use in different line service temperatures.

Figure 1-15. Dimensional gauge for bolting. (Courtesy of Lone Star Screw Co. of Houston, Inc.)

Table 1-2
Alloy Steel Machine Bolts for ANSI 150-lb
Raised Face or Flat Face Flanges, Each
with One Heavy Hex Nut

Nominal Pipe Size (in.)	Number of Machine Bolts Required	Size & Length of Machine Bolts (in.)	Wrench Size for Nut (in.)
1/2	4	1/2 × 2	7/8
3/4	4	1/2 × 2 1/4	7/8
1	4	1/2 × 2 1/4	7/8
1 1/4	4	1/2 × 2 1/2	7/8
1 1/2	4	1/2 × 2 1/2	7/8
2	4	5/8 × 2 3/4	1 1/16
2 1/2	4	5/8 × 3	1 1/16
3	4	5/8 × 3 1/4	1 1/16
3 1/2	8	5/8 × 3 1/4	1 1/16
4	8	5/8 × 3 1/4	1 1/16
5	8	3/4 × 3 1/4	1 1/4
6	8	3/4 × 3 1/2	1 1/4
8	8	3/4 × 3 3/4	1 1/4
10	12	7/8 × 4	1 7/16
12	12	7/8 × 4 1/4	1 7/16
14	12	1 × 4 1/2	1 5/8
16	16	1 × 4 3/4	1 5/8
18	16	1 1/8 × 5	1 13/16
20	20	1 1/8 × 5 1/2	1 13/16
22	20	1 1/4 × 5 3/4	2
24	20	1 1/4 × 6	2
26	24	1 1/4 × 6 1/4	2
28	28	1 1/4 × 6 1/4	2
30	28	1 1/4 × 6 1/2	2
32	28	1 1/2 × 7	2 3/8
34	32	1 1/2 × 7 1/4	2 3/8
36	32	1 1/2 × 7 1/4	2 3/8
42	36	1 1/2 × 7 3/4	2 3/8

In agreement with ANSI B. 16.5

Table 1-3
Alloy Steel Stud Bolts for ANSI 150-lb
Raised Face or Flat Face Flanges,
Each with Two Heavy Hex Nuts

Nominal Pipe Size (in.)	Number of Bolts Required	Size & Length of Stud Bolts (in.)	Wrench Size for Nuts (in.)
$1/2$	4	$1/2 \times 2^1/2$	$7/8$
$3/4$	4	$1/2 \times 2^1/2$	$7/8$
1	4	$1/2 \times 2^3/4$	$7/8$
$1^1/4$	4	$1/2 \times 2^3/4$	$7/8$
$1^1/2$	4	$1/2 \times 3$	$7/8$
2	4	$5/8 \times 3^1/4$	$1^1/16$
$2^1/2$	8	$5/8 \times 3^1/2$	$1^1/16$
3	4	$5/8 \times 3^3/4$	$1^1/16$
$3^1/2$	4	$5/8 \times 3^3/4$	$1^1/16$
4	4	$5/8 \times 3^3/4$	$1^1/16$
5	8	$3/4 \times 4$	$1^1/4$
6	8	$3/4 \times 4$	$1^1/4$
8	8	$3/4 \times 4^1/2$	$1^1/4$
10	12	$7/8 \times 4^3/4$	$1^7/16$
12	12	$7/8 \times 4^3/4$	$1^7/16$
14	12	$1 \times 5^1/4$	$1^5/8$
16	16	$1 \times 5^1/2$	$1^5/8$
18	16	$1^1/8 \times 6$	$1^{13}/16$
20	20	$1^1/8 \times 6^1/4$	$1^{13}/16$
22	20	$1^1/4 \times 6^3/4$	2
24	20	$1^1/4 \times 7$	2
26	24	$1^1/4 \times 7^1/4$	2
28	28	$1^1/4 \times 7^1/4$	2
30	28	$1^1/4 \times 7^1/2$	2
32	28	$1^1/2 \times 8^1/4$	$2^3/8$
34	32	$1^1/2 \times 8^1/4$	$2^3/8$
36	32	$1^1/2 \times 8^1/2$	$2^3/8$
42	36	$1^1/2 \times 9$	$2^3/8$

In agreement with ANSI B. 16.5

Table 1-4
Alloy Steel Stud Bolts for ANSI 300-lb
Raised Face Flanges, Each with Two
Heavy Hex Nuts

Nominal Pipe Size (in.)	Number of Bolts Required	Size & Length of Stud Bolts (in.)	Wrench Size for Nuts (in.)
½	4	½ × 2¾	⅞
¾	4	⅝ × 3	1¹/₁₆
1	4	⅝ × 3¼	1¹/₁₆
1¼	4	⅝ × 3¼	1¹/₁₆
1½	4	¾ × 3¾	1¼
2	8	⅝ × 3½	1¹/₁₆
2½	8	¾ × 4	1¼
3	8	¾ × 4¼	1¼
3½	8	¾ × 4½	1¼
4	8	¾ × 4½	1¼
5	8	¾ × 4¾	1¼
6	12	¾ × 5	1¼
8	12	⅞ × 5½	1⁷/₁₆
10	16	1 × 6¼	1⅝
12	16	1⅛ × 6¾	1¹³/₁₆
14	20	1⅛ × 7	1¹³/₁₆
16	20	1¼ × 7½	2
18	24	1¼ × 7¾	2
20	24	1¼ × 8¼	2
22	24	1½ × 9	2⅜
24	24	1½ × 9¼	2⅜
26	28	1⅝ × 10¼	2⁹/₁₆
28	28	1⅝ × 10¾	2⁹/₁₆
30	28	1¾ × 11¼	2¾
32	28	1⅞ × 12¼	2¹⁵/₁₆
34	28	1⅞ × 12½	2¹⁵/₁₆
36	32	2 × 13	3⅛

In agreement with ANSI B. 16.5

Table 1-5
Alloy Steel Stud Bolts for ANSI 400-lb
Raised Face Flanges, Each with Two
Heavy Hex Nuts

Nominal Pipe Size (in.)	Number of Bolts Required	Size & Length of Stud Bolts (in.)	Wrench Size for Nuts (in.)
$1/2$	4	$1/2 \times 3\frac{1}{4}$	$7/8$
$3/4$	4	$5/8 \times 3\frac{1}{2}$	$1\frac{1}{16}$
1	4	$5/8 \times 3\frac{3}{4}$	$1\frac{1}{16}$
$1\frac{1}{4}$	4	$5/8 \times 4$	$1\frac{1}{16}$
$1\frac{1}{2}$	4	$3/4 \times 4\frac{1}{4}$	$1\frac{1}{4}$
2	8	$5/8 \times 4\frac{1}{4}$	$1\frac{1}{16}$
$2\frac{1}{2}$	8	$3/4 \times 4\frac{3}{4}$	$1\frac{1}{4}$
3	8	$3/4 \times 5$	$1\frac{1}{4}$
$3\frac{1}{2}$	8	$7/8 \times 5\frac{1}{2}$	$1\frac{7}{16}$
4	8	$7/8 \times 5\frac{1}{2}$	$1\frac{7}{16}$
5	8	$7/8 \times 5\frac{3}{4}$	$1\frac{7}{16}$
6	12	$7/8 \times 6$	$1\frac{7}{16}$
8	12	$1 \times 6\frac{3}{4}$	$1\frac{5}{8}$
10	16	$1\frac{1}{8} \times 7\frac{1}{2}$	$1\frac{13}{16}$
12	16	$1\frac{1}{4} \times 8$	2
14	20	$1\frac{1}{4} \times 8\frac{1}{4}$	2
16	20	$1\frac{3}{8} \times 8\frac{3}{4}$	$2\frac{9}{16}$
18	24	$1\frac{3}{8} \times 9$	$2\frac{9}{16}$
20	24	$1\frac{1}{2} \times 9\frac{3}{4}$	$2\frac{3}{8}$
24	24	$1\frac{3}{4} \times 10\frac{3}{4}$	$2\frac{3}{4}$

In agreement with ANSI B. 16.5

Table 1-6
Alloy Steel Stud Bolts for ANSI 600-lb
Raised Face Flanges, Each With Two
Heavy Hex Nuts

Nominal Pipe Size (in.)	Number of Bolts Required	Size & Length of Stud Bolts (in.)	Wrench Size for Nuts (in.)
$\frac{1}{2}$	4	$\frac{1}{2} \times 3\frac{1}{4}$	$\frac{7}{8}$
$\frac{3}{4}$	4	$\frac{5}{8} \times 3\frac{1}{2}$	$1\frac{1}{16}$
1	4	$\frac{5}{8} \times 3\frac{3}{4}$	$1\frac{1}{16}$
$1\frac{1}{4}$	4	$\frac{5}{8} \times 4$	$1\frac{1}{16}$
$1\frac{1}{2}$	4	$\frac{3}{4} \times 4\frac{1}{4}$	$1\frac{1}{4}$
2	8	$\frac{5}{8} \times 4\frac{1}{4}$	$1\frac{1}{16}$
$2\frac{1}{2}$	8	$\frac{3}{4} \times 4\frac{3}{4}$	$1\frac{1}{4}$
3	8	$\frac{3}{4} \times 5$	$1\frac{1}{4}$
$3\frac{1}{2}$	8	$\frac{7}{8} \times 5\frac{1}{2}$	$1\frac{7}{16}$
4	8	$\frac{7}{8} \times 5\frac{3}{4}$	$1\frac{7}{16}$
5	8	$1 \times 6\frac{1}{2}$	$1\frac{5}{8}$
6	12	$1 \times 6\frac{3}{4}$	$1\frac{5}{8}$
8	12	$1\frac{1}{8} \times 7\frac{3}{4}$	$1\frac{13}{16}$
10	16	$1\frac{1}{4} \times 8\frac{1}{2}$	2
12	20	$1\frac{1}{4} \times 8\frac{3}{4}$	2
14	20	$1\frac{3}{8} \times 9\frac{1}{4}$	$2\frac{3}{16}$
16	20	$1\frac{1}{2} \times 10$	$2\frac{3}{8}$
18	20	$1\frac{5}{8} \times 10\frac{3}{4}$	$2\frac{9}{16}$
20	24	$1\frac{5}{8} \times 11\frac{1}{2}$	$2\frac{9}{16}$
22	24	$1\frac{3}{4} \times 12\frac{1}{4}$	$2\frac{3}{4}$
24	24	$1\frac{7}{8} \times 13$	$2\frac{15}{16}$
26	28	$1\frac{7}{8} \times 13\frac{1}{2}$	$2\frac{15}{16}$
28	28	2×14	$3\frac{1}{8}$
30	28	$2 \times 14\frac{1}{4}$	$3\frac{1}{8}$
32	28	$2\frac{1}{4} \times 15$	$3\frac{1}{2}$
36	28	$2\frac{1}{2} \times 16$	$3\frac{7}{8}$

In agreement with ANSI B. 16.5

Table 1-7
Alloy Steel Stud Bolts for ANSI 900-lb
Raised Face Flanges, Each With Two
Heavy Hex Nuts

Nominal Pipe Size (in.)	Number of Bolts Required	Size & Length of Stud Bolts (in.)	Wrench Size for Nuts (in.)
$1/2$	4	$3/4 \times 4\frac{1}{4}$	$1\frac{1}{4}$
$3/4$	4	$3/4 \times 4\frac{1}{2}$	$1\frac{1}{4}$
1	4	$7/8 \times 5$	$1\frac{7}{16}$
$1\frac{1}{4}$	4	$7/8 \times 5$	$1\frac{7}{16}$
$1\frac{1}{2}$	4	$1 \times 5\frac{1}{2}$	$1\frac{5}{8}$
2	8	$7/8 \times 5\frac{3}{4}$	$1\frac{7}{16}$
$2\frac{1}{2}$	8	$1 \times 6\frac{1}{4}$	$1\frac{5}{8}$
3	8	$7/8 \times 5\frac{3}{4}$	$1\frac{7}{16}$
4	8	$1\frac{1}{8} \times 6\frac{3}{4}$	$1\frac{13}{16}$
5	8	$1\frac{1}{4} \times 7\frac{1}{2}$	2
6	12	$1\frac{1}{8} \times 7\frac{3}{4}$	$1\frac{13}{16}$
8	12	$1\frac{3}{8} \times 8\frac{3}{4}$	$2\frac{9}{16}$
10	16	$1\frac{3}{8} \times 9\frac{1}{4}$	$2\frac{9}{16}$
12	20	$1\frac{3}{8} \times 10$	$2\frac{9}{16}$
14	20	$1\frac{1}{2} \times 10\frac{3}{4}$	$2\frac{3}{8}$
16	20	$1\frac{5}{8} \times 11\frac{1}{4}$	$2\frac{9}{16}$
18	20	$1\frac{7}{8} \times 12\frac{3}{4}$	$2\frac{15}{16}$
20	20	$2 \times 13\frac{1}{2}$	$3\frac{1}{8}$
24	20	$2\frac{1}{2} \times 17\frac{1}{4}$	$3\frac{7}{8}$

In agreement with ANSI B. 16.5

Table 1-8
Alloy Steel Stud Bolts for ANSI 1500-lb
Raised Face Flanges, Each With Two
Heavy Hex Nuts

Nominal Pipe Size (in.)	Number of Bolts Required	Size & Length of Stud Bolts (in.)	Wrench Size for Nuts (in.)
$\frac{1}{2}$	4	$\frac{3}{4} \times 4\frac{1}{4}$	$1\frac{1}{4}$
$\frac{3}{4}$	4	$\frac{3}{4} \times 4\frac{1}{2}$	$1\frac{1}{4}$
1	4	$\frac{7}{8} \times 5$	$1\frac{7}{16}$
$1\frac{1}{4}$	4	$\frac{7}{8} \times 5$	$1\frac{7}{16}$
$1\frac{1}{2}$	4	$1 \times 5\frac{1}{2}$	$1\frac{5}{8}$
2	8	$\frac{7}{8} \times 5\frac{3}{4}$	$1\frac{7}{16}$
$2\frac{1}{2}$	8	$1 \times 6\frac{1}{4}$	$1\frac{5}{8}$
3	8	$1\frac{1}{8} \times 7$	$1\frac{13}{16}$
4	8	$1\frac{1}{4} \times 7\frac{3}{4}$	2
5	8	$1\frac{1}{2} \times 9\frac{3}{4}$	$2\frac{3}{8}$
6	12	$1\frac{3}{8} \times 10\frac{1}{4}$	$2\frac{3}{16}$
8	12	$1\frac{5}{8} \times 11\frac{1}{2}$	$2\frac{9}{16}$
10	12	$1\frac{7}{8} \times 13\frac{1}{4}$	$2\frac{15}{16}$
12	16	$2 \times 14\frac{3}{4}$	$3\frac{1}{8}$
14	16	$2\frac{1}{4} \times 16$	$3\frac{1}{2}$
16	16	$2\frac{1}{2} \times 17\frac{1}{2}$	$3\frac{7}{8}$
18	16	$2\frac{3}{4} \times 19\frac{1}{2}$	$4\frac{1}{8}$
20	16	$3 \times 21\frac{1}{2}$	$4\frac{5}{8}$
24	16	$3\frac{1}{2} \times 24\frac{1}{2}$	$5\frac{3}{8}$

In agreement with ANSI B. 16.5

Table 1-9
Alloy Steel Stud Bolts for ANSI 2500-lb
Raised Face Flanges, Each With Two
Heavy Hex Nuts

Nominal Pipe Size (in.)	Number of Bolts Required	Size & Length of Stud Bolts (in.)	Wrench Size for Nuts (in.)
½	4	¾ × 5¼	1¼
¾	4	¾ × 5¼	1¼
1	4	⅞ × 5¾	1⁷/₁₆
1¼	4	1 × 6¼	1⅝
1½	4	1⅛ × 7	1¹³/₁₆
2	8	1 × 7¼	1⅝
2½	8	1⅛ × 8	1¹³/₁₆
3	8	1¼ × 9	2
4	8	1½ × 10½	2⅜
6	8	2 × 13¾	3⅛
8	12	2 × 15¼	3⅛
10	12	2½ × 19½	3⅞
12	12	2¾ × 21½	4¼

In agreement with ANSI B. 16.5

Table 1-10
Alloy Steel Stud Bolts for ANSI 150-lb
Ring Type Joint Flanges With
Two Heavy Hex Nuts Each

Nominal Pipe Size (in.)	Number of Bolts Required	Size & Length of Stud Bolts (in.)	Wrench Size for Nuts (in.)	Oval Ring Gasket R-Number
1	4	$1/2 \times 3^{1}/4$	$7/8$	R-15
$1^{1}/4$	4	$1/2 \times 3^{1}/4$	$7/8$	R-17
$1^{1}/2$	4	$1/2 \times 3^{1}/2$	$7/8$	R-19
2	4	$5/8 \times 3^{3}/4$	$1^{1}/16$	R-22
$2^{1}/2$	4	$5/8 \times 4$	$1^{1}/16$	R-25
3	4	$5/8 \times 4^{1}/4$	$1^{1}/16$	R-29
$3^{1}/2$	8	$5/8 \times 4^{1}/4$	$1^{1}/16$	R-33
4	8	$5/8 \times 4^{1}/4$	$1^{1}/16$	R-36
5	8	$3/4 \times 4^{1}/2$	$1^{1}/4$	R-40
6	8	$3/4 \times 4^{1}/2$	$1^{1}/4$	R-43
8	8	$3/4 \times 4^{3}/4$	$1^{1}/4$	R-48
10	12	$7/8 \times 5^{1}/4$	$1^{7}/16$	R-52
12	12	$7/8 \times 5^{1}/4$	$1^{7}/16$	R-56
14	12	$1 \times 5^{3}/4$	$1^{5}/8$	R-59
16	16	1×6	$1^{5}/8$	R-64
18	16	$1^{1}/8 \times 6^{1}/2$	$1^{13}/16$	R-68
20	20	$1^{1}/8 \times 6^{3}/4$	$1^{13}/16$	R-72
24	20	$1^{1}/4 \times 7^{1}/2$	2	R-76

In agreement with ANSI B. 16.5

Table 1-11
Alloy Steel Stud Bolts for ANSI 300-lb
Ring Type Joint Flanges, Each With
Two Heavy Hex Nuts

Nominal Pipe Size (in.)	Number of Bolts Required	Size & Length of Stud Bolts (in.)	Wrench Size for Nuts (in.)	Oval Ring Gasket R-Number
$1/2$	4	$1/2 \times 3$	$7/8$	R-11
$3/4$	4	$5/8 \times 3^1/2$	$1^1/16$	R-13
1	4	$5/8 \times 3^3/4$	$1^1/16$	R-16
$1^1/4$	4	$5/8 \times 3^3/4$	$1^1/16$	R-18
$1^1/2$	4	$3/4 \times 4^1/4$	$1^1/4$	R-20
2	8	$5/8 \times 4^1/4$	$1^1/16$	R-23
$2^1/2$	8	$3/4 \times 4^3/4$	$1^1/4$	R-26
3	8	$3/4 \times 5$	$1^1/4$	R-31
$3^1/2$	8	$3/4 \times 5^1/4$	$1^1/4$	R-34
4	8	$3/4 \times 5^1/4$	$1^1/4$	R-37
5	8	$3/4 \times 5^1/2$	$1^1/4$	R-41
6	12	$3/4 \times 5^3/4$	$1^1/4$	R-45
8	12	$7/8 \times 6^1/4$	$1^7/16$	R-49
10	16	1×7	$1^5/8$	R-53
12	16	$1^1/8 \times 7^1/2$	$1^{13}/16$	R-57
14	20	$1^1/8 \times 7^3/4$	$1^{13}/16$	R-61
16	20	$1^1/4 \times 8^1/4$	2	R-65
18	24	$1^1/4 \times 8^1/2$	2	R-69
20	24	$1^1/4 \times 9$	2	R-73
24	24	$1^1/2 \times 10^1/4$	$2^3/8$	R-77

In agreement with ANSI B. 16.5

Table 1-12
Alloy Steel Stud Bolts for ANSI 400-lb
Ring Type Joint Flanges, Each With Two
Heavy Hex Nuts

Nominal Pipe Size (in.)	Number of Bolts Required	Size & Length of Stud Bolts (in.)	Wrench Size for Nuts (in.)	Oval Ring Gasket R-Number
1/2	4	1/2 × 3	7/8	R-11
3/4	4	5/8 × 3 1/2	1 1/16	R-13
1	4	5/8 × 3 3/4	1 1/16	R-16
1 1/4	4	5/8 × 4	1 1/16	R-18
1 1/2	4	3/4 × 4 1/4	1 1/4	R-20
2	8	5/8 × 4 1/2	1 1/16	R-23
2 1/2	8	3/4 × 5	1 1/4	R-26
3	8	3/4 × 5 1/4	1 1/4	R-31
3 1/2	8	7/8 × 5 3/4	1 7/16	R-34
4	8	7/8 × 5 3/4	1 7/16	R-37
5	8	7/8 × 6	1 7/16	R-41
6	12	7/8 × 6 1/4	1 7/16	R-45
8	12	1 × 7	1 5/8	R-49
10	16	1 1/8 × 7 3/4	1 13/16	R-53
12	16	1 1/4 × 8 1/4	2	R-57
14	20	1 1/4 × 8 1/2	2	R-61
16	20	1 3/8 × 9	2 3/16	R-65
18	24	1 3/8 × 9 1/4	2 3/16	R-69
20	24	1 1/2 × 10	2 3/8	R-73
24	24	1 3/4 × 11 1/4	2 3/4	R-77

In agreement with ANSI B. 16.5

Table 1-13
Alloy Steel Stud Bolts for ANSI 600-lb
Ring Type Joint Flanges, Each With
Two Heavy Hex Nuts

Nominal Pipe Size (in.)	Number of Bolts Required	Size & Length of Stud Bolts (in.)	Wrench Size for Nuts (in.)	Oval Ring Gasket R-Number
½	4	½ × 3	⅞	R-11
¾	4	⅝ × 3½	1¹/₁₆	R-13
1	4	⅝ × 3¾	1¹/₁₆	R-16
1¼	4	⅝ × 4	1¹/₁₆	R-18
1½	4	¾ × 4¼	1¼	R-20
2	8	⅝ × 4½	1¹/₁₆	R-23
2½	8	¾ × 5	1¼	R-26
3	8	¾ × 5¼	1¼	R-31
3½	8	⅞ × 5¾	1⁷/₁₆	R-34
4	8	⅞ × 6	1⁷/₁₆	R-37
5	8	1 × 6¾	1⅝	R-41
6	12	1 × 7	1⅝	R-45
8	12	1⅛ × 7¾	1¹³/₁₆	R-49
10	16	1¼ × 8¾	2	R-53
12	20	1¼ × 9	2	R-57
14	20	1⅜ × 9½	2³/₁₆	R-61
16	20	1½ × 10¼	2⅜	R-65
18	20	1⅝ × 11	2⁹/₁₆	R-69
20	24	1⅝ × 11¾	2⁹/₁₆	R-73
24	24	1⅞ × 13¼	2¹⁵/₁₆	R-77
26	28	1⅞ × 14	2¹⁵/₁₆	R-93
28	28	2 × 14½	3⅛	R-94
30	28	2 × 14¾	3⅛	R-95
36	28	2½ × 16¾	3⅞	R-98

In agreement with ANSI B. 16.5

Table 1-14
Alloy Steel Stud Bolts for ANSI 900-lb
Ring Type Joint Flanges, Each With
Two Heavy Hex Nuts

Nominal Pipe Size (in.)	Number of Bolts Required	Size & Length of Stud Bolts (in.)	Wrench Size for Nuts (in.)	Oval Ring Gasket R-Number
1/2	4	3/4 × 41/4	11/4	R-12
3/4	4	3/4 × 41/2	11/4	R-14
1	4	7/8 × 5	17/16	R-16
11/4	4	7/8 × 5	17/16	R-18
11/2	4	1 × 51/2	15/8	R-20
2	8	7/8 × 53/4	17/16	R-24
21/2	8	1 × 61/4	15/8	R-27
3	8	7/8 × 6	17/16	R-31
4	8	11/8 × 7	113/16	R-37
5	8	11/4 × 73/4	2	R-41
6	12	11/8 × 73/4	113/16	R-45
8	12	13/8 × 9	23/16	R-49
10	16	13/8 × 91/2	23/16	R-53
12	20	13/8 × 101/4	23/16	R-57
14	20	11/2 × 111/4	23/8	R-62
16	20	15/8 × 113/4	29/16	R-66
18	20	17/8 × 131/2	215/16	R-70
20	20	2 × 141/4	31/8	R-74
24	20	21/2 × 173/4	37/8	R-78

In agreement with ANSI B. 16.5

Table 1-15
Alloy Steel Stud Bolts for ANSI 1500-lb
Ring Type Joint Flanges, Each With
Two Heavy Hex Nuts

Nominal Pipe Size (in.)	Number of Bolts Required	Size & Length of Stud Bolts (in.)	Wrench Size for Nuts (in.)	Oval Ring Gasket R-Number
½	4	¾ × 4¼	1¼	R-12
¾	4	¾ × 4½	1¼	R-14
1	4	⅞ × 5	1⁷/₁₆	R-16
1¼	4	⅞ × 5	1⁷/₁₆	R-18
1½	4	1 × 5½	1⅝	R-20
2	8	⅞ × 5¾	1⁷/₁₆	R-24
2½	8	1 × 6¼	1⅝	R-27
3	8	1⅛ × 7	1¹³/₁₆	R-35
4	8	1¼ × 7¾	2	R-39
5	8	1½ × 9¾	2⅜	R-44
6	12	1⅜ × 10½	2³/₁₆	R-46
8	12	1⅝ × 12	2⁹/₁₆	R-50
10	12	1⅞ × 13¾	2¹⁵/₁₆	R-54
12	16	2 × 15½	3⅛	R-58
14	16	2¼ × 17	3½	R-63
16	16	2½ × 18½	3⅞	R-67
18	16	2¾ × 20½	4¼	R-71
20	16	3 × 22½	4⅝	R-75
24	16	3½ × 25¾	5⅜	R-79

In agreement with ANSI B. 16.5

Table 1-16
Alloy Steel Stud Bolts for ANSI 2500-lb
Ring Type Joint Flanges, Each With
Two Heavy Hex Nuts

Nominal Pipe Size (in.)	Number of Bolts Required	Size & Length of Stud Bolts (in.)	Wrench Size for Nuts (in.)	Oval Ring Gasket R-Number
1/2	4	3/4 × 5 1/4	1 1/4	R-13
3/4	4	3/4 × 5 1/4	1 1/4	R-16
1	4	7/8 × 5 3/4	1 7/16	R-18
1 1/4	4	1 × 6 1/2	1 5/8	R-21
1 1/2	4	1 1/8 × 7 1/4	1 13/16	R-23
2	8	1 × 7 1/2	1 5/8	R-26
2 1/2	8	1 1/8 × 8 1/4	1 13/16	R-28
3	8	1 1/4 × 9 1/4	2	R-32
4	8	1 1/2 × 10 3/4	2 3/8	R-38
5	8	1 3/4 × 12 3/4	2 3/4	R-42
6	8	2 × 14 1/2	3 1/8	R-47
8	12	2 × 16	3 1/8	R-51
10	12	2 1/2 × 20 1/2	3 7/8	R-55
12	12	2 3/4 × 22 1/2	4 1/4	R-60

In agreement with ANSI B. 16.5

Table 1-17
Suggested Materials For Use in Various Line Service Temperatures

ASTM DESIGNATION	GRADE	SERVICE TEMPERATURE RANGE	TENSILE MIN. PSI	YIELD	REDUCTION OF AREA R/A	ELONGATION EL	ALLOY TYPE	AISI	SUITABLE NUTS (ASTM/GRADE)
ASTM A-193	B7	High Temp. from 0–480°C	125,000	100,000	50	16	Cr-Mo	(4140)	A194 Gr.2H
ASTM A-193	B16	High Temp. 0–550°C	125,000	105,000	50	18	Cr-Mo-V	(4140M)	A194 Gr.4
ASTM A-193	B7M	High Temp. 0–450°C	100,000	80,000	50	18	Cr-Mo	(4140)	A194 Gr.2HM
ASTM A-320	L7	Low Temp. −100°C	125,000	105,000	50	16	Cr-Mo	(4140)	A194 Gr.4 or A194 Gr.7
ASTM A-193	B5	Up to 815°C	100,000	80,000	50	16	5% Cr	(501)	A194 Gr.3
ASTM A-193	B6	Up to 450°C	110,000	85,000	50	16	13% Cr	(410)	A194 Gr.6
ASTM A-193	B8	Low Temp. −200°C–650°C	75,000	30,000	50	30	18% Cr-8% Ni	(304)	A194 B8
ASTM A-193	B8M	Low Temp. −200°C–750°C	75,000	30,000	50	30	16% Cr-10% Ni	(316)	A194 B8M
ASTM A-320	B8	Low Temp. −200°C–650°C	75,000	30,000	50	30	18% Cr-8% Ni	(304)	A194 B8
ASTM A-320	B8M	Low Temp. −200°C–750°C	75,000	30,000	50	30	16% Cr-10% Ni	(316)	A194 B8M

Spec	Grade	Type	Temperature	Tensile (psi)	Yield (psi)	% Elong.	% R.A.	Material	(AISI)	ASTM Equiv.
ASTM A-320	B8T		Low Temp. −200°C–650°C	75,000	30,000	50	30	17% Cr-9% Ni	(321)	A194 B8T
ASTM A-320	B8C		Low Temp. −200°C–650°C	75,000	30,000	50	30	17% Cr-9% Ni	(347)	A194 B8C
ASTM A-193	B8 Class 2		Low Temp. −200°C–650°C	125,000	100,000	35	12	18% Cr-8% Ni	(304)	A194 B8
ASTM A-453		660	High Temp. to 750°C	130,000	85,000	18	15	High Iron Superalloy	660 (A286)	A453 660
ASTM A-564 (cond. 900)		630	Med. Temp. up to 316°C	190,000	170,000	40	10	Precipitation Hardening STST	630 (17-4)	A564 630
(cond. 1100)				140,000	115,000	45	14			
ASTM B-408		800–800H	High Temp. 540°C–815°C	80,000	35,000	25		Incoloy		ASTM B408
ASTM B-164	405		High Temp. to 815°C	85,000	50,000	15		Monel		
ASTM B-164	K500		High Temp. 650°C–423°C / Low Temp. 650°C–253°C	100,000	70,000	35		Monel		
600, 625, 718, X750			High Temp. Range from 815°C–1090°C	100,000 / 170,000	80,000 / 115,000	30 / 30	30 / 30	Nickel Superalloy (Inconel) Nickel Superalloy (Hastelloy)		
B, C, X			High Temp. to 850°C							

TEMPERATURE CONVERSIONS

°C	°F
− 423	− 793
− 200	− 392
− 100	− 212
316	601
450	842
480	896
540	1004
550	1022
650	1202
750	1382
815	1499
850	1562

(Courtesy Lone Star Screw Company of Houston, Inc.)

2

ANSI
BUTTWELD
FITTINGS

ANSI buttweld fittings are used to change direction or join parts of a piping system. Mastering the names of the various shapes is not too difficult, because the number of shapes is limited.

Figure 2-1 shows the identification markings that are required on all fittings. The 90-degree long radius elbow is marked with the size and schedule number, the material grade, and the heat code symbol, also known as the laboratory control number.

Types of Buttweld Fittings

Elbows

The elbow is the most commonly used fitting, and the long radius elbow is probably the most commonly used elbow. The short radius elbow is used in systems with tight spaces, such as offshore and skid units.

Figure 2-2 shows the 90-degree long radius elbow next to a 90-degree short radius elbow. Remember that a 90-

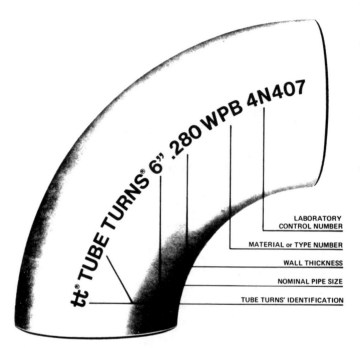

Figure 2-1. Identification markings. (Courtesy of Tube Turns Technologies, Inc.)

degree long radius elbow has a center-to-face dimension of one and one-half times the size of the elbow. The center-to-face dimension of a short radius elbow is the same as the size of the elbow; e.g. a 2-in. short radius elbow has a center-to-face dimension of 2 ins. For other dimensions, see Table 2-1.

Figure 2-2. 90-degree elbows, long and short radius. (Courtesy of Hackney, Inc., a division of Trinity Industries.)

Figure 2-3. 90-degree reducing elbow, long radius. (Courtesy of Hackney, Inc., a division of Trinity Industries.)

Reducing Elbows

The 90-degree reducing elbow is used to change direction and reduce the flow in piping systems. See Figure 2-3.

43

Table 2-1
Seamless Welding Fittings
(Courtesy of Taylor-Bonney Division, a Gulf + Western Manufacturing Company.)

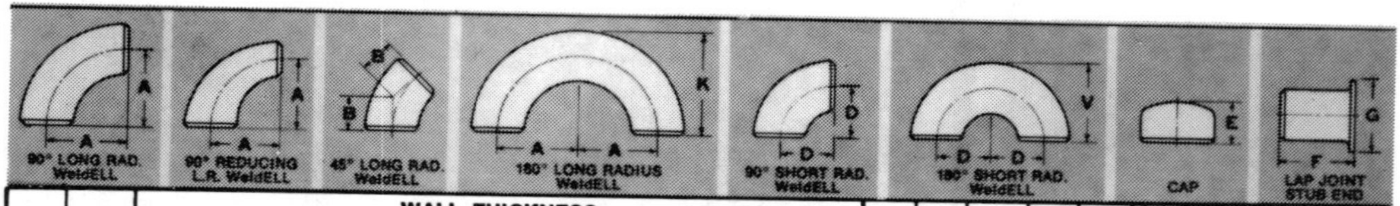

Nom. Pipe Size	Pipe O.D.	WALL THICKNESS ① Light Wall	Sch 20	Sch 30	② Std	Sch 40	Sch 60	③ X-Stg	Sch 80	Sch 100	Sch 120	Sch 140	Sch 160	XX-Stg	A	B	K	D	V	⑦ E	F ASA	F MSS	G	Nom. Pipe Size
½	.840	.083	...	...	.109	.109	...	.147	.147	...	...	...	.188	.294	1½	⅝	1⅞	...	...	1	3	2	1⅜	½
¾	1.050	.083	...	...	.113	.113	...	.154	.154	...	...	...	.219	.308	1⅛	7/16	1 11/16	...	...	1	3	2	1 11/16	¾
1	1.315	.109	...	...	.133	.133	...	.179	.179	...	...	...	.250	.358	1½	⅞	2 3/16	1	1⅝	1½	4	2	2	1
1¼	1.660	.109	...	...	.140	.140	...	.191	.191	...	...	...	.250	.382	1⅞	1	2¾	1¼	2 1/16	1½	4	2	2½	1¼
1½	1.900	.109	...	...	.145	.145	...	.200	.200	...	...	...	.281	.400	2¼	1⅛	3¼	1½	2 7/16	1½	4	2	2⅞	1½
2	2.375	.109	...	...	.154	.154	...	.218	.218	...	...	...	.344	.436	3	1⅜	4 3/16	2	3 3/16	1½	6	2½	3⅝	2
2½	2.875	.120	...	...	.203	.203	...	.276	.276	...	...	...	.375	.552	3¾	1¾	5 3/16	2½	3 15/16	1½	6	2½	4⅛	2½
3	3.500	.120	...	...	.216	.216	...	.300	.300	...	...	...	.438	.600	4½	2	6¼	3	4¾	2	6	2½	5	3
3½	4.000	.120	...	...	.226	.226	...	.318	.318	...	...	...	...	.636	5¼	2¼	7¼	3½	5½	2½	6	3	5½	3½
4	4.500	.120	...	...	.237	.237	...	.337	.337	...	.438	...	.531	.674	6	2½	8¼	4	6¼	2½	6	3	6 3/16	4
5	5.563	.134	...	...	.258	.258	...	.375	.375	...	.500	...	.625	.750	7½	3⅛	10 5/16	5	7¾	3	8	3	7 5/16	5
6	6.625	.134	...	...	.280	.280	...	.432	.432	...	.562	...	.719	.864	9	3¾	12 5/16	6	9 5/16	3½	8	3½	8½	6
8	8.625	.148	.250	.277	.322	.322	.406	.500	.500	.594	.719	.812	.906	.875	12	5	16 5/16	8	12 5/16	4	8	4	10⅝	8
10	10.750	.165	.250	.307	.365	.365	.500	.500	.594	.719	.844	1.000	1.125	1.000	15	6¼	20⅜	10	15⅜	5	10	5	12¾	10
12	12.750	.180	.250	.330	.375	.406	.562	.500	.688	.844	1.000	1.125	1.312	1.000	18	7½	24⅜	12	18⅜	6	10	6	15	12
14	14.000	.250	.312	.375	.375	.438	.594	.500	.750	.938	1.094	1.250	1.406	...	21	8¾	28	14	21	6½	12	...	16¼	14
16	16.000	.250	.312	.375	.375	.500	.656	.500	.844	1.031	1.219	1.438	1.594	...	24	10	32	16	24	7	12	...	18½	16
18	18.000	.250	.312	.438	.375	.562	.750	.500	.938	1.156	1.375	1.562	1.781	...	27	11¼	36	18	27	8	12	...	21	18
20	20.000	.250	.375	.500	.375	.594	.812	.500	1.031	1.281	1.500	1.750	1.969	...	30	12½	40	20	30	9	12	...	23	20
24	24.000	.250	.375	.562	.375	.688	.969	.500	1.219	1.531	1.812	2.062	2.344	...	36	15	48	24	36	10½	12	...	27¼	24
30	30.000	.312	.500	.625	.375	...	...	.500	...	...	...	...	...	...	45	18½	60	30	45	10½	...	...	...	30
36 ④	36.000	.312	.500	.625	.375	.750	...	.500	...	...	...	...	...	...	54	22¼	...	36	54	10½	...	...	...	36 ④
42 ④	42.000	...	...	...	.375	...	...	.500	...	...	...	...	...	...	63	26	...	42	...	12	...	...	...	42 ④
48 ④	48.000	...	...	...	.375	...	...	.500	...	...	...	...	...	...	72	29¾	...	48	...	13½	...	...	...	48 ④

(Continued on next page)

Table 2-1 continued

STRAIGHT TEE

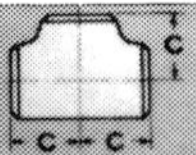

Nom. Pipe Size	Outlet	C	M	H
¾	¾	1⅛	. . .	. . .
	½	1⅛	1⅛	1½
1	1	1½	. . .	. . .
	¾	1½	1½	2
	½	1½	1½	2
1¼	1¼	1⅞	. . .	. . .
	1	1⅞	1⅞	2
	¾	1⅞	1⅞	2
	½	1⅞	1⅞	2
1½	1½	2¼	. . .	. . .
	1¼	2¼	2¼	2½
	1	2¼	2¼	2½
	¾	2¼	2¼	2½
	½	2¼	2¼	2½
2	2	2½	. . .	. . .
	1½	2½	2⅜	3
	1¼	2½	2¼	3
	1	2½	2	3
	¾	2½	1¾	3
2½	2½	3	. . .	. . .
	2	3	2¾	3½
	1½	3	2⅝	3½
	1¼	3	2½	3½
	1	3	2¼	3½
3	3	3⅜	. . .	. . .
	2½	3⅜	3¼	3½
	2	3⅜	3	3½
	1½	3⅜	2⅞	3½
	1¼	3⅜	2¾	3½

REDUCING TEE

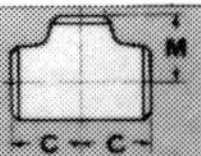

Nom. Pipe Size	Outlet	C	M	H
3½	3½	3¾	. . .	. . .
	3	3¾	3⅝	4
	2½	3¾	3½	4
	2	3¾	3¼	4
	1½	3¾	3⅛	4
4	4	4⅛	. . .	. . .
	3½	4⅛	4	4
	3	4⅛	3⅞	4
	2½	4⅛	3¾	4
	2	4⅛	3½	4
	1½	4⅛	3⅜	4
5	5	4⅞	. . .	. . .
	4	4⅞	4⅝	5
	3½	4⅞	4½	5
	3	4⅞	4⅜	5
	2½	4⅞	4¼	5
	2	4⅞	4⅛	5
6	6	5⅝	. . .	. . .
	5	5⅝	5⅜	5½
	4	5⅝	5⅛	5½
	3½	5⅝	5	5½
	3	5⅝	4⅞	5½
	2½	5⅝	4¾	5½
8	8	7	. . .	. . .
	6	7	6⅝	6
	5	7	6⅜	6
	4	7	6⅛	6
	3½	7	6	6

CONCENTRIC REDUCER

Nom. Pipe Size	Outlet	C	M	H
10	10	8½	. . .	. . .
	8	8½	8	7
	6	8½	7⅝	7
	5	8½	7½	7
	4	8½	7¼	7
12	12	10	. . .	. . .
	10	10	9½	8
	8	10	9	8
	6	10	8⅝	8
	5	10	8½	8
14	14	11	. . .	. . .
	12	11	10⅝	13
	10	11	10⅛	13
	8	11	9¾	13
	6	11	9⅜	13
16	16	12	. . .	. . .
	14	12	12	14
	12	12	11⅝	14
	10	12	11⅛	14
	8	12	10¾	14
	6	12	10⅜	14
18	18	13½	. . .	. . .
	16	13½	13	15
	14	13½	13	15
	12	13½	12⅝	15
	10	13½	12⅛	15
	8	13½	11¾	15

ECCENTRIC REDUCER

Nom. Pipe Size	Outlet	C	M	H
20	20	15	. . .	. . .
	18	15	14½	20
	16	15	14	20
	14	15	14	20
	12	15	13⅝	20
	10	15	13⅛	20
	8	15	12¾	20
24 ④	24	17	. . .	. . .
	20	17	17	20
	18	17	16½	20
	16	17	16	20
	14	17	16	20
	12	17	15⅝	20
	10	17	15⅛	20
30 ④	30	22	. . .	. . .
	24	22	21	24
	20	22	20	24
	18	22	19½	24
	16	22	19	24
	14	22	19	24
36 ④	36	26½	. . .	. . .
	30	26½	25	24
	24	26½	24	24
	20	26½	23	24
	18	26½	22½	24
	16	26½	22	24
42 ④	42	30	28	. . .
	36	30	28	24
	30	30	28	24
	24	30	26	24
	20	30	26	24
48	48	35	33	. . .
	42	35	32	28
	36	35	31	28
	30	35	30	28

NOTES:
① Light Wall thicknesses are identical to stainless steel Schedule 10S in sizes thru 12", and to Schedule 10 in sizes 14" and larger.
② Standard Wall thicknesses are identical to stainless steel Schedule 40S in sizes thru 12".
③ Extra Strong Wall thicknesses are identical to stainless steel Schedule 80S in sizes thru 12".
④ May be of welded pipe, x-rayed and stress-relieved.
⑤ Other types, sizes and thicknesses of fittings on application.
⑥ Stocked in carbon steel and a variety of other metals and alloys.
⑦ See ANSI B16.9 for cap lengths when wall thicknesses are greater than x-stg.

Figure 2-4. 45-degree elbow, long radius. (Courtesy of Hackney, Inc., a division of Trinity Industries.)

45-degree Elbows

Figure 2-4 shows a 45-degree long radius elbow. These elbows are used for partial changes in direction of the line. The 45-degree elbows are sometimes trimmed to a lesser degree when required, such as 37 degrees.

180-degree Returns

The return is used for direction changes of 180-degrees, thus avoiding the use of two 90-degree elbows. Figure 2-5 depicts a long and short radius 180-degree return.

Tees

A tee is a branched connection to the main flow, and can be either straight or reducing, as shown in Figure

Figure 2-5. 180-degree returns, long and short radius. (Courtesy of Hackney, Inc., a division of Trinity Industries.)

Figure 2-6. Straight and reducing tees. (Courtesy of Hackney, Inc., a division of Trinity Industries.)

2-6. The reducing outlet can be specified on any branch, and Figure 2-7 shows the correct descriptions used for reducing tees and other fittings.

Crosses

Straight or reducing crosses are seldom used in systems, except where space requirements dictate it. Figure 2-8 shows a straight cross. Crosses are made in sizes of 12-in. and smaller.

How to Read Reducing Fittings

Elbows

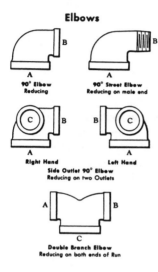

90° Elbow
Reducing

90° Street Elbow
Reducing on male end

Right Hand

Left Hand

Side Outlet 90° Elbow
Reducing on two Outlets

Double Branch Elbow
Reducing on both ends of Run

True "Y"

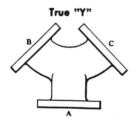

To assist the user in "reading" reducing fittings, a variety of types most commonly required for piping systems are illustrated on this page. In these illustrations, each opening of the fitting is identified with a letter which indicates the sequence to be followed in reading the size of the fitting.

In designating the outlets of reducing fittings, the openings should be read in the order indicated by the sequence of the letters "A", "B", "C", and "D". On side outlet reducing fittings, the size of the side outlet is named last.

For example: A Cross having one end of the run and one outlet reduced is designated as:

$$
\begin{array}{cccc}
A & B & C & D \\
2\frac{1}{2} & \times\ 1\frac{1}{4} & \times\ 2\frac{1}{2} & \times\ 1\frac{1}{2}
\end{array}
$$

Simply name the largest opening first and then name the other openings in the order indicated.

Note: Although all but one of the illustrations are of screwed fittings, the same rules apply to the "reading" of reducing flanged, welding, solder-joint, and other types of fittings.

Service Tee

Service Tee
Reducing on male end only

48

(Continued on next page)

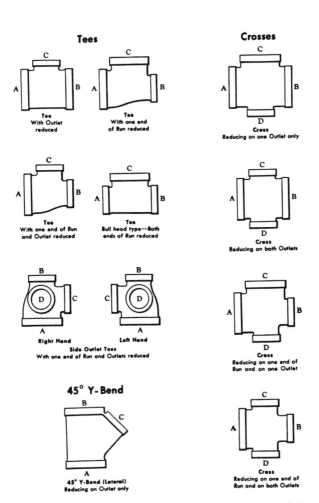

Figure 2-7. How to read reducing fittings. (Courtesy of Crane Co.)

49

Figure 2-8. Weld cross. (Courtesy of Hackney, Inc., a division of Trinity Industries.)

Reducers

Eccentric and concentric reducers, illustrated in Figure 2-9, are used to reduce a line to a smaller size. Very few eccentric reducers are used in piping systems, therefore it is not difficult to tell which is which. The concentric reducer has an inlet and outlet that are on a center line.

The eccentric reducer has an off-center outlet, and is flat on one side. The eccentric reducer fits flush against a wall, ceiling, or floor to give greater pipe support to the line.

Figure 2-9. Concentric and eccentric weld reducers. (Courtesy of Hackney, Inc., a division of Trinity Industries.)

50

Lap Joint Stub Ends

The stub end is used in lines requiring quick disconnection. See Figure 2-10. The lap forms a gasket surface that replaces the gasket surface of a flange, and are mated with a lap joint flange. (Refer to Figure 1-4).

Stub ends should not be confused with stub-ins, the latter being one pipe stubbed into another pipe and welded. See Figure 2-11.

Figure 2-10. Lap joint stub end. (Courtesy of Hackney, Inc., a division of Trinity Industries.)

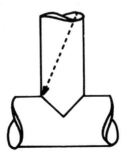

Figure 2-11. A stub-in.

51

Figure 2-12. Weld pipe cap. (Courtesy of Hackney, Inc., a division of Trinity Industries.)

Caps

Pipe caps are used to block off the end of a line by welding it to the pipe. Caps should never be stored in a position to trap rain water or sand. See Figure 2-12.

Special Buttweld Fittings

Pipe Saddles

The saddle, as shown in Figure 2-13, is used to reinforce a junction of pipe or fitting in a line. After a nipple has been welded into a line, the saddle is placed over the outlet, and welded to both the outlet and the line.

Figure 2-13. Pipe saddle. (Courtesy of Hackney, Inc., a division of Trinity Industries.)

Figure 2-14. 45-degree lateral. (Courtesy of Hackney, Inc., a division of Trinity Industries.)

Laterals

Figure 2-14 shows a 45-degree lateral. Low-pressure applications are about the only time laterals will be used.

SCRAPER BAR TEES

Figure 2-15. Scraper bar tee. (Courtesy of Hackney, Inc., a division of Trinity Industries.)

Scraper Bar Tee

Figure 2-15 shows that bars have been fabricated inside the outlet of a tee. The bars limit the direction a pipeline scraper (or "Pig"), can travel inside a pipeline.

Material Control

Protecting Weld Fittings

Store weld fittings in a position so that water or sand will not collect inside them.

54

Large diameter fittings can be stored and protected with end covers, either plastic or metal. Fittings 1½ ins. and smaller should be stored inside, away from the elements.

Fittings can be stacked in layers with or without dunnage. Metal to metal contact will not harm adjacent fittings.

In corrosive areas, spray the fittings with a specified preservative to prevent rust.

Do not throw or dump the fittings from containers. Permanent damage to a fitting's beveled ends may result.

When you are receiving fittings from a vendor or from another area, always check each and every fitting for damage and markings per the purchase order specifications.

Mixed Schedule Fittings

It is very common to have fittings of a mixed schedule match different pipe wall thicknesses. When the fittings are machined to a lower pipe schedule, the process is called "taper boring." Two examples would be:

1. One 12-in. 90-degree weld elbow, schedule 60 long radius, taper bore each end to extra strong.
2. A 12-in. by 6-in. concentric reducer, schedule 120, taper bore the 12-in. end to schedule 100, the 6-in. end to schedule 80.

When you receive such fittings from vendors, mark each fitting with positive markings for future identification. (The vendor should have already marked the fittings for you).

Do not store altered fittings with regular sized fittings. A wrong fitting installed by accident in a system could prove disastrous.

Backing Rings

Figure 2-16 shows rings that are sometimes used in piping systems under severe service conditions. One type ring is grooved with knockoff spacer pins.

Figure 2-16. Grove type welding ring with knock-off spacer pins. (Courtesy of Tube Turns Technologies, Inc.)

56

Figure 2-17. Flat type and ridge type welding rings. (Courtesy of Tube Turns Technologies, Inc.)

The backup rings are inserted in the adjoining ends of pipes that are to be buttwelded. The rings prevent spatter and metal icicles from forming inside the pipe. The ring becomes a permanent part of the piping system. Figure 2-17 shows two other type backing rings—the flat and ridge types.

Branch Olet Connections

There are many reputable manufacturers of Olet fittings used for branch connections. For description and illustration purposes, the Bonney Forge fittings are described here, along with their registered trade names for the fittings.

Interchangeability

Table 2-2 shows the interchangeability and size consolidation of the Olet fittings. The correct descriptions of

Table 2-2
Interchangeability or Consolidation of Sizes
(Courtesy of Bonney Forge.)

Standard Weight and Extra Strong Weldolet △ △
3000# Thredolet & 3000# Sockolet

RUN SIZES	OUTLET SIZE INCHES														
	1/8	1/4	3/8	1/2	3/4	1	1¼	1½	2	2½	3	3½	4	5	6
	3/8	3/8	1/2	1/2	3/4	1/4	1½	1½	2	3	3	3½	4	5	6
	1-3/4	1-3/4	1-3/4	3/4	1-3/4	1/2	1½	2	2½	3½	3½	4	5	6	8
	1-3/4	1-3/4	2½-1¼	1½-1¼	1⅛-1¼	1-3/4	2	2½-3	3	4	4	5	6	8	10
	2¾-1¼	2¾-1¼	36-3	2½-2	2½-3	2-3/3	3-3/3	3½-3	4-3½	5	5	8	8	10	14-12
	36-3	36-3	Flat	8-3	5-3	5-4	5-4	5-4	6	6	6	8	10-12	12	16
	Flat	Flat		36-10	12-6	10-6	18-10	8-6	10-8	12-10	8	10	14-12	14	18
	△	△	3/4-½	Flat	36-14	36-16	Flat	24-14	18-12	18-14	10-12	14-12	20-16	18-16	22-20
			36-1		Flat	Flat		Flat	36-20	36-20	14-12	20-16	22	22-20	28-24
			Flat						Flat	Flat	20-16	36-24	36-24	28-24	36-30
											36-24	Flat	Flat	36-30	Flat
											Flat			Flat	

Outlet Sizes: 8, 10, 12, 14, 16, 18, 20, 24, 26, 30 order to specific size combination

△ For 1/8 & 1/4 Weldolets—no run size variation is available. Order by outlet size only.
△ Size consolidation for Weldolets only.

6000# Thredolet & 6000# Sockolet

RUN SIZES	OUTLET SIZE INCHES							
	1/4	3/8	1/2	3/4	1	1¼	1½	2
	1/2	1/2	1-3/4	1	1½-1¼	2	2	2½
	1-3/4	3/4	2-1¼	2½-1¼	2½-2	2½-2	2½-3	3
	2½-1¼	1	6-2½	10-3	10-3	3½-3	3½-3	4
	36-3	1½-1¼	36-8	36-12	36-12	8-4	8-6	6
	Flat	2½	Flat	Flat	Flat	36-10	18-10	10-8
		8-3				Flat	36-20	20-12
		36-10					Flat	36-24
		Flat						Flat

(Continued on next page)

58

Table 2-2 continued

Schedule 160 and XXS Weldolet

RUN SIZES	OUTLET SIZE INCHES								
	½	¾	1	1¼	1½	2	2½	3	3½
	½-¾	1-¾	1	1½-1¼	1½	2	2½-2	3	3½-2
	1¼-3¾	2-1¼	2½-1¼	2½-2	2½-2	2½-2	3½-3	3½-2	4
	36-1½	6-2½	10-3	10-3	3½-3	3½-3	5-4	4	5
	Flat	36-8	36-12	36-12	8-4	5-4	8-6	5	6
		Flat	Flat	Flat	20-10	8-6	12-10	6	8
					22	18-10	18-14	8	10
					36-24	36-20	36-20	12-10	14-12
					Flat	Flat	Flat	14	20-16
								16	22
								18	36-24
								20	Flat
								22	
								36-24	
								Flat	

Outlet Sizes 4, 5, 6, 8, 10, 12 order to specific size combination

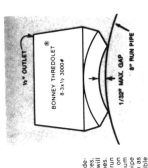

BONNEY THREDOLET ®
8-3×½ 3000#
½" OUTLET
1/32" MAX. GAP
8" RUN PIPE

THE WELDOLET MEANS REDUCED INVENTORY

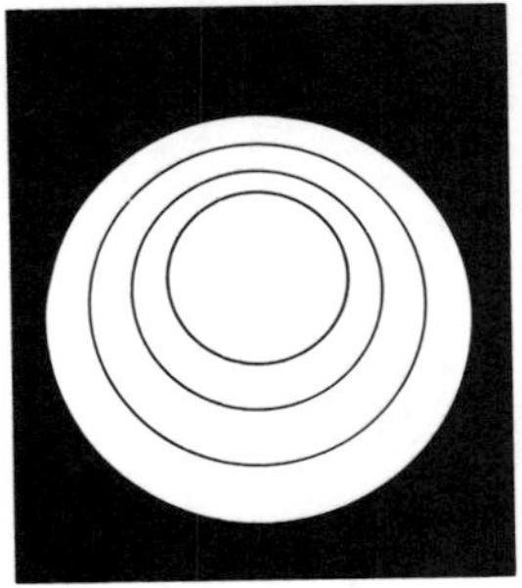

The chart above outlines the full range of Thredolet, Sockolet and Weldolet size consolidation. This chart has been devised and the fittings designed to substantially reduce warehouse inventory. All fittings are manufactured and marked as shown on the chart. All outlet sizes not listed on chart should be ordered to specific run pipe size.

HOW IT WORKS

Each outlet size indicated on chart is designed to fit a number of run pipe sizes. e.g... the ½" fitting marked 8-3 × ½ will fit 3", 3½", 4", 5", 6", and 8" run pipes. When this ½" fitting is placed on a 3" run pipe, it will fit perfectly. When placed on an 8" run pipe, there will be a maximum gap of 1/32" between the top of the run pipe and the base of the fitting at the crotch as shown on sketch. This gap is negligible when welding.

59

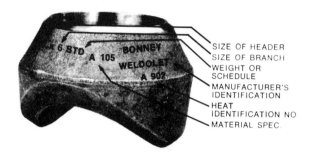

SIZE OF HEADER
SIZE OF BRANCH
WEIGHT OR SCHEDULE
MANUFACTURER'S IDENTIFICATION
HEAT IDENTIFICATION NO
MATERIAL SPEC.

BRANCH & HEADER
SAME WEIGHT OR SCHEDULE

SCHEDULE OF HEADER
SCHEDULE OF BRANCH
HEAT NUMBER

BRANCH & HEADER
DIFFERENT WEIGHT OR SCHEDULE

Figure 2-18. Markings on branch connections. (Courtesy of Bonney Forge.)

60

the Olets are shown in the table, and are suitable for use in your materials accounting system.

Notice in Table 2-2, the run size numbers—36-3, 8-3, 12-6, etc. These and the other numbers fit run sizes from the high number down through the low number. The system is further explained in Table 2-2.

Figure 2-18 illustrates the identification markings that are required on Olet fittings.

Thredolets

This is a fitting that is buttwelded on the run of pipe and has a threaded outlet. It is widely used on all projects. See Figure 2-19.

Sockolet

This fitting is exactly the same as the Thredolet except it has a socket weld outlet. See Figure 2-19.

Sweepolet®

Resembling a saddle, this fitting is strong enough to support the branch line being buttwelded to it. See Figure 2-19. Sweepolets will not be used too often on your projects.

Elbolet®

This fitting is welded to a 90-degree elbow to form an outlet. See Figure 2-19. Elbolets are available with

WELDOLET

THREDOLET

SOCKOLET

ELBOLET

SWEEPOLET

Figure 2-19. Bonney Forge fittings. (Courtesy of Bonney Forge.)

threaded, socket weld, and buttweld outlets. Table 2-3 lists the sizes of elbolets and the correct descriptions.

Flatolet®

This fitting is used on flat surface areas, such as weld caps and heads.

Table 2-3
Bonney Forge Elbolets®
(Courtesy of Bonney Forge.)

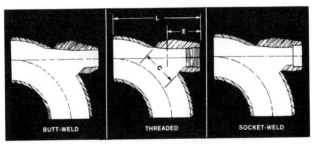

BUTT-WELD   THREADED   SOCKET-WELD

°NOMINAL ELBOW SIZE INCHES	OUTLET SIZE INCHES	DIMENSIONS			
		3000± THREADED & SOCKET WELD Std. & XS Butt-Weld		6000± THREADED & SOCKET WELD Sch. 160 & XXS Butt-Weld	
		C	E	C	E
36-1¼ 36-1¼	¼ ⅜	1½ 1½	1¹⁹⁄₃₂ 1¹⁹⁄₃₂	1½ 1½	1¹⁹⁄₃₂ 1¹⁹⁄₃₂
36-1¼ 36-1¼	½ ¾	1½ 1²³⁄₃₂	1¹⁹⁄₃₂ 1⅞	1²³⁄₃₂ 2¼	1⅞ 2¼
36-2 36-2	1 1¼	2¼ 2⅞	2¼ 2½	2⅞ 3⅛	2½ 2¹¹⁄₁₆
36-2 36-3	1½ 2	3⅛ 4³⁄₁₆	2¹¹⁄₁₆ 3¼	4³⁄₁₆	3¼
ORDER TO SPECIFIC ELBOW SIZES	°°2½ °°3 °°4 °°6 °°8 °°10 °°12	4³⁄₁₆ 5¹⁄₁₆ 6⅝ 9⅜ 13⁵⁄₁₆ 17⁵⁄₃₂ 19⅝	3¼ 3⅞ 4¹³⁄₁₆ 6½ 8¹⁄₁₆ 10⅜ 11⅛		

Socket Weld Reducing Inserts

Socket Weld
Steel Weld Couplet

Threaded Steel
Weld Couplet

NIPOLET:
Plain End or Threaded

LATROLET

Figure 2-20. Bonney Forge Nipolet®, Latrolet®, and couplets; reducing inserts by Henry Vogt Machine Company. (Courtesy of Bonney Forge and Henry Vogt Machine Company.)

Storing Olets

All threaded sizes of Olets should be stored inside. Fittings 2 ins. and larger with weld ends may be stored outdoors. Small weld fittings should be stored indoors. Cardboard bin boxes are an excellent storage method to separate the smaller Olets inside of warehouse shelving for easy inventory and issue.

Other Olets

Some of the other Olet fittings are Nipolets®, Latrolets®, couplets, reducing inserts, and etc. See Figure 2-20.

3

REFINERY
PIPE

Standard pipe is widely used in the oil and gas industries, and is manufactured to ASTM standards (ANSI B36.10). Pipe charts, such as the one in Table 3-1, and careful attention to purchase order descriptions when shipping or receiving pipe help achieve accurate results. Therefore, a description of piping definitions and how various types are manufactured follow.

Pipe Size

In pipe of any given size, the variations in wall thickness do not affect the outside dimension (OD), just the inside dimension (ID). For example, 12-in. nominal pipe has the same OD whether the wall thickness is 0.375 in. or 0.500 in. (Refer to Table 3-1 for wall thickness of pipe).

(Text continued on page 70)

Table 3-1
Pipe Chart
(Courtesy of Tioga Pipe Supply Company)

NOMINAL PIPE SIZE INCHES	OUTSIDE DIAMETER INCHES	I.P.S.	SCHEDULE	WALL INCHES	INSIDE DIAMETER INCHES	WT/FT POUNDS
⅛	.405		10S	.049	.307	.1863
		40	40S Std.	.068	.269	.2447
		80	80S Ex. Hvy	.095	.215	.3145
¼	.540		10S	.065	.410	.3297
		40	40S Std.	.088	.364	.4248
		80	80S Ex. Hvy	.119	.302	.5351
⅜	.675		10S	.065	.545	.4235
		40	40S Std.	.091	.493	.5676
		80	80S Ex. Hvy	.126	.423	.7388
½	.840		5S	.065	.710	.5383
			10S	.083	.674	.6710
		40	40S Std.	.109	.622	.8510
		80	80S Ex Hvy	.147	.546	1.088
		160		.188	.466	1.309
			XX Hvy	.294	.252	1.714
¾	1.050		5S	.065	.920	.6838
			10S	.083	.884	.8572
		40	40S Std.	.113	.824	1.131
		80	80S Ex Hvy	.154	.742	1.474
		160		.219	.614	1.944
			XX Hvy	.308	.434	2.441
1	1.315		5S	.065	1.185	.8678
			10S	.109	1.097	1.404
		40	40S Std.	.133	1.049	1.679
		80	80S Ex Hvy	.179	.957	2.172
		160		.250	.815	2.844
			XX Hvy	.358	.599	3.659

(Continued on next page)

Table 3-1 continued

NOMINAL PIPE SIZE INCHES	OUTSIDE DIAMETER INCHES	I.P.S.	SCHEDULE	WALL INCHES	INSIDE DIAMETER INCHES	WT/FT POUNDS
1¼	1.660		5S	.065	1.530	1.107
			10S	.109	1.442	1.806
		40	40S Std	.140	1.380	2.273
		80	80S Ex Hvy	.191	1.278	2.997
		160		.250	1.160	3.765
			XX Hvy	.382	.896	5.214
1½	1.900		5S	.065	1.770	1.274
			10S	.109	1.682	2.085
		40	40S Std	.145	1.610	2.718
		80	80S Ex Hvy	.200	1.500	3.631
		160		.281	1.338	4.859
			XX Hvy	.400	1.100	6.408
2	2.375		5S	.065	2.245	1.604
			10S	.109	2.157	2.638
		40	40S Std	.154	2.067	3.653
		80	80S Ex Hvy	.218	1.939	5.022
		160		.344	1.689	7.462
			XX Hvy	.436	1.503	9.029
2½	2.875		5S	.083	2.709	2.475
			10S	.120	2.635	3.531
		40	40S Std	.203	2.469	5.793
		80	80S Ex Hvy	.276	2.323	7.661
		160		.375	2.125	10.01
			XX Hvy	.552	1.771	13.69
3	3.500		5S	.083	3.334	3.029
			10S	.120	3.260	4.332
		40	40S Std.	.216	3.068	7.576
		80	80S Ex Hvy.	.300	2.900	10.25
		160		.438	2.624	14.32
			XX Hvy	.600	2.300	18.58

(Continued on next page)

Table 3-1 continued

NOMINAL PIPE SIZE INCHES	OUTSIDE DIAMETER INCHES	I.P.S.	SCHEDULE	WALL INCHES	INSIDE DIAMETER INCHES	WT/FT POUNDS
3½	4.000	5	5S	083	3.834	3.472
		10	10S	120	3.760	4.973
		40	40S Std	226	3.548	9.109
		80	80S Ex Hvy	318	3.364	12.50
			XX Hvy	636	2.728	22.85
4	4.500		5S	083	4.334	3.915
			10S	120	4.260	5.613
		40	40S Std	237	4.026	10.79
		80	80S Ex Hvy	337	3.826	14.98
		120		438	3.624	19.00
		160		531	3.438	22.51
			XX Hvy	674	3.152	27.54
4½	5.00		40 Std	247	4.506	12.53
			80 Ex Hvy	355	4.290	17.61
			XX Hvy	710	3.580	32.43
5	5.563		5S	109	5.345	6.349
			10S	134	5.295	7.770
		40	40S Std	258	5.047	14.62
		80	80S Ex Hvy	375	4.813	20.78
		120		500	4.563	27.04
		160		625	4.313	32.96
			XX Hvy	750	4.063	38.55
6	6.625		5S	109	6.407	7.585
			10S	134	6.357	9.289
		40	40S Std	280	6.065	18.97
		80	80S Ex Hvy	432	5.761	28.57
		120		562	5.491	36.39
		160		719	5.189	45.35
			XX Hvy	864	4.897	53.16

(Continued on next page)

Table 3-1 continued

NOMINAL PIPE SIZE INCHES	OUTSIDE DIAMETER INCHES	I.P.S.	SCHEDULE	WALL INCHES	INSIDE DIAMETER INCHES	WT/FT POUNDS
7	7.625	40	Std.	.301	7.023	23.57
		80	Ex. Hvy.	.500	6.625	38.05
			XX Hvy.	.875	5.875	63.08
8	8.625		5S	.109	8.407	9.914
			10S	.148	8.329	13.40
		20		.250	8.125	22.36
		30		.277	8.071	24.70
		40	40S Std.	.322	7.981	28.55
		60		.406	7.813	35.64
		80	80S Ex. Hvy.	.500	7.625	43.39
		100		.594	7.439	50.95
		120		.719	7.189	60.71
		140		.812	7.001	67.76
			XX Hvy.	.875	6.875	72.42
		160		.906	6.813	74.69
9	9.625	40	Std.	.342	8.941	33.90
		80	Ex. Hvy.	.500	8.625	48.72
			XX Hvy.	.875	7.875	81.77
10	10.750		5S	.134	10.482	15.19
			10S	.165	10.420	18.70
		20		.250	10.250	28.04
		30		.307	10.136	34.24
		40	40S Std.	.365	10.020	40.48
		60	80S Ex. Hvy.	.500	9.750	54.74
		80		.594	9.564	64.43
		100		.719	9.314	77.03
		120		.844	9.064	89.29
		140		1.000	8.750	104.13
		160		1.125	8.500	115.64
11	11.750	40	Std.	.375	11.000	45.55
		80	Ex. Hvy.	.500	10.750	60.07
			XX Hvy.	.875	10.000	101.63

(Continued on next page)

Table 3-1 continued

NOMINAL PIPE SIZE INCHES	OUTSIDE DIAMETER INCHES	I.P.S.	SCHEDULE	WALL INCHES	INSIDE DIAMETER INCHES	WT/FT POUNDS
12	12.750		5S	.165	12.420	22.18
			10S	.180	12.390	24.20
		20		.250	12.250	33.38
		30		.330	12.090	43.77
			40S Std.	.375	12.000	49.56
		40		.406	11.938	53.52
			80S Ex. Hvy.	.500	11.750	65.42
		60		.562	11.626	73.15
		80		.688	11.376	88.63
		100		.844	11.064	107.32
		120		1.000	10.750	125.49
		140		1.125	10.500	139.67
		160		1.312	10.126	160.27
14	14.000	10		.250	13.500	36.71
		20		.312	13.376	45.61
		30	Std.	.375	13.250	54.57
		40		.438	13.124	63.44
			Ex. Hvy.	.500	13.000	72.09
		60		.594	12.814	85.05
		80		.750	12.500	106.13
		100		.938	12.126	130.85
		120		1.094	11.814	150.9
		140		1.250	11.500	170.21
		160		1.406	11.188	189.1
16	16.000	10		.250	15.500	42.05
		20		.312	15.376	52.27
		30	Std.	.375	15.250	62.58
		40	Ex. Hvy.	.500	15.000	82.77
		60		.656	14.688	107.5
		80		.844	14.314	136.61
		100		1.031	13.938	164.82
		120		1.219	13.564	192.43
		140		1.438	13.124	223.64
		160		1.594	12.814	245.25

(Continued on next page)

Table 3-1 continued

NOMINAL PIPE SIZE INCHES	OUTSIDE DIAMETER INCHES	I.P.S. SCHEDULE		WALL INCHES	INSIDE DIAMETER INCHES	WT/FT POUNDS
18	18.000	10		250	17.500	47.39
		20		312	17.376	58.94
			Std.	375	17.250	70.59
		30		438	17.124	82.15
			Ex. Hvy	500	17.000	93.45
		40		562	16.876	104.67
		60		750	16.500	138.17
		80		938	16.126	170.92
		100		1.156	15.688	207.96
		120		1.375	15.250	244.14
		140		1.562	14.876	274.22
		160		1.781	14.438	308.5
20	20.000	10		250	19.500	52.73
		20	Std	375	19.250	78.60
		30	Ex. Hvy	500	19.000	104.13
		40		594	18.814	123.11
		60		812	18.376	166.4
		80		1.031	17.938	208.87
		100		1.281	17.438	256.1
		120		1.500	17.000	296.37
		140		1.750	16.500	341.09
		160		1.969	16.064	379.17
22	22.000	10		250	21.500	58.07
		20	Std.	375	21.250	86.61
		30	X Hvy	500	21.000	114.81
		60		875	20.250	197.41
		80		1.125	19.750	250.81
		100		1.375	19.250	302.88
		120		1.625	18.750	353.61
		140		1.875	18.250	403.0
		160		2.125	17.750	451.06
24	24.000	10		250	23.500	63.41
		20		375	23.250	94.62
			Std.			
			Ex. Hvy	500	23.000	125.49

72

(Continued on next page)

Table 3-1 continued

NOMINAL PIPE SIZE INCHES	OUTSIDE DIAMETER INCHES	I.P.S.	SCHEDULE	WALL INCHES	INSIDE DIAMETER INCHES	WT/FT POUNDS
		30		.562	22.876	140.68
		40		.688	22.626	171.29
		60		.969	22.064	238.35
		80		1.219	21.564	296.58
		100		1.531	20.938	367.39
		120		1.812	20.376	429.39
		140		2.062	19.876	483.1
		160		2.344	19.314	542.13
26	26.000	10		.312	25.376	85.60
			Std	.375	25.250	102.63
		20	X Hvy.	.500	25.000	136.17
28	28.000	10		.312	27.376	92.26
			Std	.375	27.250	110.64
		20		.500	27.000	146.85
		30		.625	26.750	182.73
30	30.000	10		.312	29.376	98.93
			Std	.375	29.250	118.65
		20	Ex Hvy	.500	29.000	157.53
		30		.625	28.750	196.08
32	32.000	10		.312	31.376	105.59
			Std	.375	31.250	126.66
		20		.500	31.000	168.21
		30		.625	30.750	209.43
		40		.688	30.624	230.08
34	34.000	10		.312	33.376	112.25
			Std	.375	33.250	134.67
		20		.500	33.000	178.89
		30		.625	32.750	222.78
		40		.688	32.624	244.77
36	36.000	10		.312	35.375	118.92
			Std	.375	35.250	142.68
			Ex Hvy	.500	35.000	189.57

(Continued on next page)

Table 3-1 continued

NOMINAL PIPE SIZE INCHES	OUTSIDE DIAMETER INCHES	I.P.S.	SCHEDULE	WALL INCHES	INSIDE DIAMETER INCHES	WT/FT POUNDS
42	42.000		Std.	.375	41.250	166.71
		20	X Hvy.	.500	41.000	221.61
		30		.625	40.750	276.18
		40		.750	40.500	330.41
48	48.000		Std.	.375	47.250	190.74
			X Hvy.	.500	47.000	253.65

Pipe Length

Pipe is supplied and referred to as single random, double random, longer than double random, and cut lengths.

Single random pipe length is usually 18–22 ft threaded and coupled (T&C), and 18–25 ft plain end (PE).

Double random pipe lengths average 38–40 feet.

Cut lengths are made to order within $\pm 1/8$-in. Some pipe is available in about 80-ft lengths.

The major manufacturers of pipe offer brochures on their process of manufacturing pipe. The following descriptions are based upon vendor literature and specifications.

74

Seamless Pipe

This type of pipe is made by heating billets and advancing them over a piercer point. The pipe then passes through a series of rolls where it is formed to a true round and sized to exact requirements.

Electric Weld

Coils or rolls of flat steel are fed to a forming section that transforms the flat strip of steel into a round pipe section. A high-frequency welder heats the edges of the strip to 2,600°F at the fusion point. Pressure rollers then squeeze the heated edges together to form a fusion weld.

Double Submerged Arc Weld

Flat plate is used to make large-diameter pipe (20-in.– 44 in.) in double random lengths. The plate is rolled and pressed into an "O" shape, then welded at the edges both inside and outside. The pipe is then expanded to the final diameter.

Continuous Weld

Coiled skelp (skelp is semi-finished coils of steel plate used specifically for making pipe), is fed into a flattener, and welded to the trailing end of a preceding coil, thus forming a continuous strip of skelp. The skelp travels through a furnace where it is heated to 2,600°F and then bent into an oval by form rollers. It then proceeds through a welding stand where the heat in the skelp and pressure exerted by the rolls forms the weld. The pipe is stretched to a desired OD and ID, and cut to lengths. (Couplings if ordered for any size pipe will be hand tight only.)

Pipe Specifications

ASTM A-120

Sizes ⅛-in. to 16-ins., standard weight, extra strong, and double extra strong (Std. Wt., XS, XXS). The specification covers black and hot-dipped galvanized welded and seamless average wall pipe for use in steam, gas, and air lines.

Markings. Rolled, stamped or stenciled on each length of pipe: the brand name, ASTM A-120, and the length of the pipe. In case of bundled pipe, markings will appear on a tag attached to each bundle. Table 3-2 shows a bundling schedule.

ASTM A-53

Sizes ⅛-in. to 26-ins., standard weight, extra strong, and double extra strong, ANSI schedules 10 through 160 (see Table 3-1 for ANSI pipe schedules). The specification covers seamless and welded black and hot-dipped

Table 3-2 Bundling Schedule

Nominal Pipe Size (in.)	Number pieces per Bundle	Standard Weight Pipe		Extra Strong Pipe	
		Total length (ft)	Total weight (lbs)	Total length (ft)	Total weight (lbs)
1/8	30	630	151	630	195
1/4	24	504	212	504	272
3/8	18	378	215	378	280
1/2	12	252	214	252	275
3/4	7	147	166	147	216
1	5	105	176	105	228
1 1/4	3	63	144	63	189
1 1/2	3	63	172	63	229

galvanized average wall pipe for conveying oil, water, gas, and petroleum products.

Markings. Rolled, stamped or stenciled with brand name, kind, schedule, length of pipe, and type of steel used. In case of bundles, markings will appear on a bundle tag.

ASTM A-106

Sizes 1/8 to 26-ins., ANSI schedules to 160. The specification covers seamless carbon steel average wall pipe for high-temperature service.

Markings. Rolled, stamped or stenciled with brand name, type such as ASTM A-106A, A-106B, A-106C (the A, B, C, indicate tensile strengths and yield point designations), the test pressure, and length of pipe. In case of bundles, the markings will appear on a bundle tag.

API-5L

Sizes $\frac{1}{8}$-in. to 48-ins., standard weight through double extra strong. The specification covers welded and seamless pipe suitable for use in conveying oil, water, and gas.

Markings. Paint stenciled with brand name, the API monogram, size, grade, steel process, type of steel, length, and weight per foot on pipe 4-ins. and larger. In case of bundles, the markings will be on the bundle tag. Couplings, if ordered, will be hand tight.

Storing Pipe

Step 1—Pipe Racks

Figure 3-1 shows a pipe rack made by using 12 × 12-in. timbers. The rack has been assigned a number for materials accounting purposes. Do not store pipe directly on the ground. If rack materials are not available, then use the pipe itself by preparing a rack from the pipe with a few boards under each end.

Step 2—Layers

Form the first layer of pipe with one end straight, and other joints straight across the rack. Secure the stack by nailing wooden blocks to the sills, against the side of the pipe on the inside edges (see Figure 3-1).

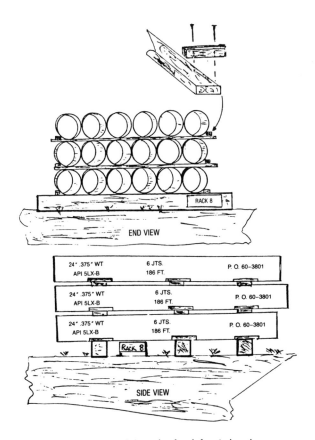

END VIEW

| 24" .375" WT | 6 JTS. | |
| API 5LX-B | 186 FT. | P. O. 60–3801 |

| 24" .375" WT | 6 JTS. | |
| API 5LX-B | 186 FT. | P. O. 60–3801 |

| 24" .375" WT | 6 JTS. | |
| API 5LX-B | 186 FT. | P. O. 60–3801 |

RACK 8

SIDE VIEW

Figure 3-1. Schematic of rack for storing pipe.

Step 3—Measure

Tally each joint of pipe in the layer. Use a paint stick or suitable marker to mark each joint according to length, size, schedule, and purchase order item number.

Total the footage on the layer of pipe, and then mark the total footage and number of joints on the outside pipe for future inventory purposes. Apply color codes to pipe at this time if applicable.

Step 4—Dunnage

Apply sufficient dunnage of the same thickness across the pipe with wooden blocks nailed to one side. Stack the next layer of pipe directly over the first layer with the straight ends in line with each other. Then follow steps 2, 3, and 4.

Continue to follow the steps until the rack is considered full by the supervisor.

Rules for Storing Pipe

1. Do not mix pipe sizes and schedules on the same pipe rack.
2. Keep the pipe storage area clean to prevent accidents.
3. Do not crowd the storage areas. Leave room for large trucks and cranes.
4. Make a physical count of the pipe on a weekly or monthly basis to verify your materials accounting records as correct.

5. Always measure pipe within tenths of an inch. Measure the entire length of pipes, including couplings and threads.

Calculations to Use

If the outside diameter (OD) and the wall thickness of a pipe (t) are known, then you may calculate the weight per foot with the following equation:

Weight per foot $= 10.68 \times (OD - t) \times t$

Example: What is the weight per foot of a 3-in. pipe with a .216-in. wall thickness and an OD of 3.500 ins.? Using the equation,

Weight per foot $= 10.68 \times (3.500 - .216) \times .216$
$= 7.58 \text{ lbs/ft}$

Another method to determine weight per foot of pipe where the outside diameter and wall thickness are known is called the Baiamonte plate method. It is based on a square foot of plate 1 inch thick weighing 40.833 lbs, and uses the following equation:

Weight per foot $= 40.833 \times \left(\dfrac{OD - t}{12}\right) \times \pi \times t$

Example: What is the weight per foot of an 8-in. pipe with a wall thickness of .322 in.? Table 3-1

shows that an 8-in. pipe has an OD of 8.625 ins. So, using the equation,

$$\text{Weight per foot} = 40.833 \times \left(\frac{8.625 - .322}{12} \right)$$

$$\times 3.1416 \times t$$

$$= 28.58 \text{ lbs/ft}$$

4

API FLANGES

The difference between API and ANSI flanges is the material from which they are fabricated and the higher working pressure at which API flanges may be operated.

API flanges are manufactured primarily for use with oil industry high-strength tubular goods. The API 6A and ANSI B. 16.5 flanges are similar dimensionally (see Table 4-1), but they cannot be interconnected without affecting the overall working pressure rating.

Another difference is the through-bore nominal size designation, such as $1^{13}/_{16}$ and $2^{1}/_{16}$, for 6B flanges in place of old nominal sizes, such as $1^{1}/_{2}$-inch, and 2-inch, for consistency with 6BX flange size designations. (See Tables 4-2 and 4-3.)

Some API flanges with casing or tubing threads have hub lengths greater than required for ANSI flanges.

The new bore size designations for API flanges or clamp type connectors, will take several years to become a routine part of the materials vocabulary. Therefore, for reference, Tables 4-2 and 4-3 contain a complete

Table 4-1
API vs. ANSI Flange Dimensions

Flange Type	Pressure Class Rating		Nominal Size Range (in.)		Old API Nominal size (in.)
	ANSI	API	ANSI	API	
Weldneck	600	2000	1/2–24	1 13/16 to 11	1 1/2 to 10
	900	3000	1/2–24	1 13/16 to 11	1 1/2 to 10
	1500	5000	1/2–24	1 13/16 to 11	1 1/2 to 10
Blind, Threaded, & Integral	600	2000	1/2–24	1 13/16 to 21 1/4	1 1/2 to 20
	900	3000	1/2–24	1 13/16 to 20 3/4	1 1/2 to 20
	1500	5000	1/2–24	1 13/16 to 11	1 1/2 to 10

Table 4-2
API Bore Sizes and Matching Tubular Goods Sizes
for 2,000, 3,000 and 5,000-lb psi Flanges
or 5,000 lb psi Clamp Type Connectors

New API Bore Sizes for Flanges and Hubs	Old Nominal Flange Size (in.)	Line Pipe Nominal Size (in.)	Tubing Outside Diameter (in.)	Casing Outside Diameter (in.)
1 13/16	1 1/2	1 1/2	1.660 & 1.900	
2 1/16	2	2	1.600 thru 2 3/8	
2 9/16	2 1/2	2 1/2	2 7/8	
3 1/8	3	3	3 1/2	
4 1/16	4	4	4 & 4 1/2	
7 1/16	6	6		4 1/2
9	8	8		4 1/2 thru 7
11	10	10		7 5/8 & 8 5/8
13 5/8	12	12		9 5/8 & 10 3/4
13 5/8	13 5/8	—		11 3/4 & 13 3/8
16 3/4	16	16		11 3/4 & 13 3/8
16 3/4	16 3/4	—		16
21 1/4	20	20		16
20 3/4	20	20		20
				20

Table 4-3
API Bore Sizes and Matching Tubular Goods Sizes
for 10,000, 15,000 and 20,000 lb psi Flanges
and 10,000 lb. psi Clamp Type Connectors

New API Bore Sizes for Flanges and Hubs (in.)	Tubing Outside Diameter (in.)	Casing Outside Diameter (in.)
1 $11/16$	1.900	
1 $13/16$	2.063	
2 $1/16$	2 $3/8$	
2 $9/16$	2 $7/8$	
3 $1/16$	3 $1/2$	
4 $1/16$	4 & 4 $1/2$	4 $1/2$
7 $1/16$		4 $1/2$ through 7
9		7 $5/8$ & 8 $5/8$
11		8 $5/8$ & 9 $5/8$
13 $5/8$		10 $3/4$ & 11 $3/4$
16 $3/4$		16
18 $3/4$		18 $5/8$
21 $1/4$		20

list of the new bore size designations, the old nominal size designations, and the matching tubular goods sizes for use with API flanges or clamp type connectors.

Bore diameter of API flanges should be the same inside diameter as the pipe to be used.

API flanges are marked with the API monogram (a registered trademark), size, pressure rating, ring gasket size, bore, manufacturer, and a heat number. Some API flanges are marked with the manufacturers' part or assembly numbers. Consult their individual catalogs for respective part numbers and descriptions.

Figure 4-1. API threaded flange. (Courtesy of National Supply Company.)

Figure 4-2. API Type 6BX weldneck flange. (Courtesy of National Supply Company.)

API Type 6B and 6BX Flanges
2,000–20,000 psi Maximum Working Pressures

Figure 4-1 illustrates a *threaded flange.* Threaded flanges do not have a bore schedule, but do have a description of the type of thread, such as casing, tubing, or line pipe. The flange illustrated was made by National Supply Company, and will service 15,000 lb. WOG pressure.

Figure 4-2 shows a *weld neck flange* manufactured by the same company, which will service 15,000 lbs. WOG pressure. This type flange does have a bore schedule.

Tables 4-4 through 4-13, indicate the size range, bolt requirements, wrench size for the bolts, and the ring gasket number required for API type 6B and 6BX flanges.

(Text continued on page 88)

Table 4-4
API Type 6B Flanges for 2000 psi Maximum Working Pressure

Nominal Size and Bore of Flange (in.)	Old Nominal Size of Flange (in.)	Number of Bolts	Size and Length of Bolts (in.)	Wrench Size for Bolts (in.)	Ring Number R or RX
$1^{13}/_{16}$	$1^{1}/_{2}$	4	$^{3}/_{4} \times 4^{1}/_{4}$	$1^{1}/_{4}$	20
$2^{1}/_{16}$	2	8	$^{5}/_{8} \times 4^{1}/_{2}$	$1^{1}/_{16}$	23
$2^{9}/_{16}$	$2^{1}/_{2}$	8	$^{3}/_{4} \times 5$	$1^{1}/_{4}$	26
$3^{1}/_{8}$	3	8	$^{3}/_{4} \times 5^{1}/_{4}$	$1^{1}/_{4}$	31
$4^{1}/_{16}$	4	8	$^{7}/_{8} \times 6$	$1^{7}/_{16}$	37
$5^{1}/_{8}$	5	8	$1 \times 6^{3}/_{4}$	$1^{5}/_{8}$	41
$7^{1}/_{16}$	6	12	1×7	$1^{5}/_{8}$	45
9	8	12	$1^{1}/_{8} \times 8$	$1^{13}/_{16}$	49
11	10	16	$1^{1}/_{4} \times 8^{3}/_{4}$	2	53
$13^{5}/_{8}$	12	20	$1^{1}/_{4} \times 9$	2	57
$16^{3}/_{4}$	16	20	$1^{1}/_{2} \times 10^{1}/_{4}$	$2^{3}/_{8}$	65
$17^{3}/_{4}$	18	20	$1^{5}/_{8} \times 11$	$2^{9}/_{16}$	69
$21^{1}/_{4}$	20	24	$1^{5}/_{8} \times 11^{3}/_{4}$	$2^{9}/_{16}$	73

Table 4-5
API Type 6B Flanges for 3000 psi Maximum Working Pressure

Nominal Size and Bore of Flange (in.)	Old Nominal Size of Flange (in.)	Number of Bolts	Size and Length of Bolts (in.)	Wrench Size for Bolts (in.)	Ring Number R or RX
$1\frac{13}{16}$	$1\frac{1}{2}$	4	$1 \times 5\frac{1}{2}$	$1\frac{5}{8}$	20
$2\frac{1}{16}$	2	8	$\frac{7}{8} \times 6$	$1\frac{7}{16}$	24
$2\frac{9}{16}$	$2\frac{1}{2}$	8	$1 \times 6\frac{1}{2}$	$1\frac{5}{8}$	27
$3\frac{1}{8}$	3	8	$\frac{7}{8} \times 6$	$1\frac{7}{16}$	31
$4\frac{1}{16}$	4	8	$1\frac{1}{8} \times 7$	$1\frac{13}{16}$	37
$5\frac{1}{8}$	5	8	$1\frac{1}{4} \times 7\frac{3}{4}$	2	41
$7\frac{1}{16}$	6	12	$1\frac{1}{8} \times 8$	$1\frac{13}{16}$	45
9	8	12	$1\frac{3}{8} \times 9$	$2\frac{3}{16}$	49
11	10	16	$1\frac{3}{8} \times 9\frac{1}{2}$	$2\frac{3}{16}$	53
$13\frac{5}{8}$	12	20	$1\frac{3}{8} \times 10\frac{1}{4}$	$2\frac{3}{16}$	57
$16\frac{3}{4}$	16	20	$1\frac{5}{8} \times 11\frac{3}{4}$	$2\frac{9}{16}$	66
$17\frac{3}{4}$	18	20	$1\frac{7}{8} \times 13\frac{3}{4}$	$2\frac{15}{16}$	70
$20\frac{3}{4}$	20	20	$2 \times 14\frac{1}{2}$	$3\frac{1}{8}$	74

Table 4-6
API Type 6B Flanges for 5,000 psi Maximum Working Pressure

Nominal Size and Bore of Flange (in.)	Old Nominal Size of Flange (in.)	Number of Bolts	Size and Length of Bolts (in.)	Wrench Size for Bolts (in.)	Ring Number R or RX
$1\frac{13}{16}$	$1\frac{1}{2}$	4	$1 \times 5\frac{1}{2}$	$1\frac{5}{8}$	20
$2\frac{1}{16}$	2	8	$\frac{7}{8} \times 6$	$1\frac{7}{16}$	24
$2\frac{9}{16}$	$2\frac{1}{2}$	8	$1 \times 6\frac{1}{2}$	$1\frac{5}{8}$	27
$3\frac{1}{8}$	3	8	$1\frac{1}{8} \times 7\frac{1}{4}$	$1\frac{13}{16}$	35
$4\frac{1}{16}$	4	8	$1\frac{1}{4} \times 8$	2	39
$5\frac{1}{8}$	5	8	$1\frac{1}{2} \times 10$	$2\frac{3}{8}$	44
$7\frac{1}{16}$	6	12	$1\frac{3}{8} \times 10\frac{3}{4}$	$2\frac{3}{16}$	46
9	8	12	$1\frac{5}{8} \times 12$	$2\frac{9}{16}$	50
11	10	12	$1\frac{7}{8} \times 13\frac{3}{4}$	$2\frac{15}{16}$	54

Table 4-7
API Type 6BX Weldneck Flanges for 10,000 psi
Maximum Working Pressure

Nominal Size and Bore of Flange (in.)	Number of Bolts	Size and Length of Bolts (in.)	Wrench Size for Bolts (in.)	Ring Number	Weight of Flange (lbs)
$1^{11}/_{16}$	8	$^3/_4 \times 5$	$1^1/_4$	BX-150	22
$1^{13}/_{16}$	8	$^3/_4 \times 5$	$1^1/_4$	BX-151	24
$2^1/_{16}$	8	$^3/_4 \times 5^1/_4$	$1^1/_4$	BX-152	38
$2^9/_{16}$	8	$^7/_8 \times 6$	$1^7/_{16}$	BX-153	38
$3^1/_{16}$	8	$1 \times 6^3/_4$	$1^5/_8$	BX-154	52
$4^1/_{16}$	8	$1^1/_8 \times 8$	$1^{13}/_{16}$	BX-155	66
$5^1/_8$	12	$1^1/_8 \times 8^3/_4$	$1^{13}/_{16}$	BX-169	120
$7^1/_{16}$	12	$1^1/_2 \times 11^1/_4$	$2^3/_8$	BX-156	340
9	16	$1^1/_2 \times 13$	$2^3/_8$	BX-157	550
11	16	$1^3/_4 \times 15$	$2^3/_4$	BX-158	810
$13^5/_8$	20	$1^7/_8 \times 17^1/_4$	$2^{15}/_{16}$	BX-159	970
$16^3/_4$	24	$1^7/_8 \times 17^1/_2$	$2^{15}/_{16}$	BX-162	1400

Table 4-8
API Type 6BX Weldneck Flanges for 15,000 psi
Maximum Working Pressure

Nominal Size and Bore of Flange (in.)	Number of Bolts	Size and Length of Bolts (in.)	Wrench Size for Bolts (in.)	Ring Number	Weight of Flange (lbs)
$1^{11}/_{16}$	8	$^3/_4 \times 5^1/_4$	$1^1/_4$	BX-150	22
$1^{13}/_{16}$	8	$^7/_8 \times 5^1/_2$	$1^7/_{16}$	BX-151	28
$2^1/_{16}$	8	$^7/_8 \times 6$	$1^7/_{16}$	BX-152	34
$2^9/_{16}$	8	$1 \times 6^3/_4$	$1^5/_8$	BX-153	34
$3^1/_{16}$	8	$1^1/_8 \times 7^1/_2$	$1^{13}/_{16}$	BX-154	64
$4^1/_{16}$	8	$1^3/_8 \times 9^1/_4$	$2^3/_{16}$	BX-155	154
$7^1/_{16}$	16	$1^1/_2 \times 12^3/_4$	$2^3/_8$	BX-156	440

Table 4-9
API Type 6BX Weldneck Flanges for 20,000 psi
Maximum Working Pressure

Nominal Size and Bore of Flange (in.)	Number of Bolts	Size and Length of Bolts (in.)	Wrench Size for Bolts (in.)	Ring Number	Weight of Flange (lbs)
$1\,^{13}/_{16}$	8	$1 \times 7\,^{1}/_{2}$	$1\,^{5}/_{8}$	BX-151	35
$2\,^{1}/_{16}$	8	$1\,^{1}/_{8} \times 8\,^{1}/_{4}$	$1\,^{13}/_{16}$	BX-152	52
$2\,^{9}/_{16}$	8	$1\,^{1}/_{4} \times 9\,^{1}/_{4}$	2	BX-153	65
$3\,^{1}/_{16}$	8	$1\,^{3}/_{8} \times 10$	$2\,^{3}/_{16}$	BX-154	140
$4\,^{1}/_{16}$	8	$1\,^{3}/_{4} \times 12\,^{1}/_{4}$	$2\,^{3}/_{4}$	BX-155	270
$7\,^{1}/_{16}$	16	$2 \times 17\,^{1}/_{2}$	$3\,^{1}/_{8}$	BX-156	620

Table 4-10
API Type 6BX Integral Flanges for 5,000 psi
Maximum Working Pressure

Nominal Size and Bore of Flange (in.)	Number of Bolts	Size and Length of Bolts (in.)	Wrench Size for Bolts (in.)	Ring Number
$13\,^{5}/_{8}$	16	$1\,^{5}/_{8} \times 12\,^{1}/_{2}$	$2\,^{9}/_{16}$	BX-160
$16\,^{3}/_{4}$	16	$1\,^{7}/_{8} \times 14\,^{1}/_{2}$	$2\,^{15}/_{16}$	BX-162
$18\,^{3}/_{4}$	20	$2 \times 17\,^{1}/_{2}$	$3\,^{1}/_{8}$	BX-163
$21\,^{1}/_{4}$	24	$2 \times 18\,^{3}/_{4}$	$3\,^{1}/_{8}$	BX-165

Table 4-11
API Type 6BX Integral Flanges for 10,000 psi
Maximum Working Pressure

Nominal Size and Bore of Flange (in.)	Number of Bolts	Size and Length of Bolts (in.)	Wrench Size for Bolts (in.)	Ring Number
$1^{11}/_{16}$	8	$^3/_4 \times 5$	$1^1/_4$	BX-150
$1^{13}/_{16}$	8	$^3/_4 \times 5$	$1^1/_4$	BX-151
$2^1/_{16}$	8	$^3/_4 \times 5^1/_4$	$1^1/_4$	BX-152
$2^9/_{16}$	8	$^7/_8 \times 6$	$1^7/_{16}$	BX-153
$3^1/_{16}$	8	$1 \times 6^3/_4$	$1^5/_8$	BX-154
$4^1/_{16}$	8	$1^1/_8 \times 8$	$1^{13}/_{16}$	BX-155
$5^1/_8$	12	$1^1/_8 \times 8^3/_4$	$1^{13}/_{16}$	BX-169
$7^1/_{16}$	12	$1^1/_2 \times 11^1/_4$	$2^3/_8$	BX-156
9	16	$1^1/_2 \times 13$	$2^3/_8$	BX-157
11	16	$1^3/_4 \times 15$	$2^3/_4$	BX-158
$13^5/_8$	20	$1^7/_8 \times 17^1/_4$	$2^{15}/_{16}$	BX-159
$16^3/_4$	24	$1^7/_8 \times 17^1/_2$	$2^{15}/_{16}$	BX-162
$18^3/_4$	24	$2^1/_4 \times 22^1/_2$	$3^1/_2$	BX-164
$21^1/_4$	24	$2^1/_2 \times 24^1/_2$	$3^7/_8$	BX-166

Table 4-12
API Type 6BX Integral Flanges for 15,000 psi
Maximum Working Pressure

Nominal Size and Bore of Flange (in.)	Number of Bolts	Size and Length of Bolts (in.)	Wrench Size for Bolts (in.)	Ring Number
$1^{11}/_{16}$	8	$^3/_4 \times 5^1/_4$	$1^1/_4$	BX-150
$1^{13}/_{16}$	8	$^7/_8 \times 5^1/_2$	$1^7/_{16}$	BX-151
$2^1/_{16}$	8	$^7/_8 \times 6$	$1^7/_{16}$	BX-152
$2^9/_{16}$	8	$1 \times 6^3/_4$	$1^5/_8$	BX-153
$3^1/_{16}$	8	$1^1/_8 \times 7^1/_2$	$1^{13}/_{16}$	BX-154
$4^1/_{16}$	8	$1^3/_8 \times 9^1/_4$	$2^3/_{16}$	BX-155
$7^1/_{16}$	16	$1^1/_2 \times 12^3/_4$	$2^3/_8$	BX-156
9	16	$1^7/_8 \times 15^3/_4$	$2^{15}/_{16}$	BX-157
11	20	$2 \times 19^1/_4$	$3^1/_8$	BX-158

Table 4-13
API Type 6BX Integral Flanges for 20,000 psi
Maximum Working Pressure

Nominal Size and Bore of Flange (in.)	Number of Bolts	Size and Length of Bolts (in.)	Wrench Size for Bolts (in.)	Ring Number
1^{13}/$_{16}$	8	1 × 7^{1}/$_{2}$	1^{5}/$_{8}$	BX-151
2^{1}/$_{16}$	8	1^{1}/$_{8}$ × 8^{1}/$_{4}$	1^{13}/$_{16}$	BX-152
2^{9}/$_{16}$	8	1^{1}/$_{4}$ × 9^{1}/$_{4}$	2	BX-153
3^{1}/$_{16}$	8	1^{3}/$_{8}$ × 10	2^{3}/$_{16}$	BX-154
4^{1}/$_{16}$	8	1^{3}/$_{4}$ × 12^{1}/$_{4}$	2^{3}/$_{4}$	BX-155
7^{1}/$_{16}$	16	2 × 17^{1}/$_{2}$	3^{1}/$_{8}$	BX-156

Table 4-14
Recommended 6BX Flange Bolt Torque

Bolt Size	Torque (ft-lbs)
^{3}/$_{4}$ -10 UNC	200
^{7}/$_{8}$ -9 UNC	325
1 -8 UNC	475
1^{1}/$_{8}$-8 UN	600
1^{3}/$_{8}$-8 UN	1200
1^{1}/$_{2}$-8 UN	1400
1^{5}/$_{8}$-8 UN	1700
1^{3}/$_{4}$-8 UN	2040
1^{7}/$_{8}$-8 UN	3220
2 -8 UN	3850

Table 4-14 is the recommended bolt torque requirements and wrench size required on API type 6BX flanges.

Figure 4-3 shows a Type 6BX weld neck flange with a transition piece. The transition piece, when ordered, is applied by the manufacturer due to the difficulty of field welding and heat treating at the jobsite. The API monogram will not appear on the transition piece. Lengths of transition pieces will vary.

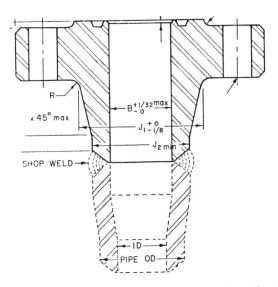

Figure 4-3. API Type 6BX weldneck flange with translation piece. (Courtesy of American Petroleum Institute.)

Weights of some API flanges and clamp type connectors appear in the tables for use as shipping weights only and are not a part of the API specifications.

Bolting and Ring Gaskets for API Flanges

API type 6B flanges require an R or RX ring gasket. API type 6BX flanges require a BX ring gasket. R and RX ring gaskets are interchangeable with each other. BX gaskets are not interchangeable with R and RX ring gaskets. (See Tables 4-20 and 4-21.)

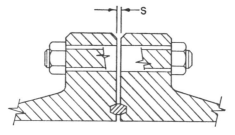

API TYPE 6B FLANGE

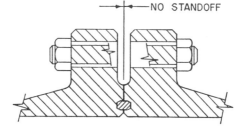

API TYPE 6BX FLANGE

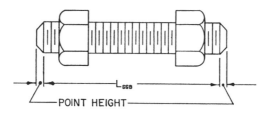

STUD BOLT WITH NUTS

Figure 4-4. API flange standoff difference and bolt lengths. (Courtesy of American Petroleum Institute.)

Bolt lengths have been calculated to accommodate the standoff difference shown in Figure 4-4 for 6B and 6BX flanges. Also illustrated in Figure 4-4, are the point heights for stud bolts. Point heights are not included in the calculations for stud bolt lengths.

The lengths for point heights in inches are as follows:

Bolt Diameter	Maximum Point Height
$1/2$ to $7/8$-in.	$1/8$-in.
$7/8$ to $1 1/8$	$3/16$
$1 1/8$ to $1 5/8$	$1/4$
$1 5/8$ to $1 7/8$	$5/16$
$1 7/8$ to $2 1/4$	$3/8$

Dual Completion Flanges

Figure 4-5 shows a typical 5,000-lb psi-working-pressure segmented flange used for dual completions. Table 4-15 shows the bolt requirements, API ring gasket number, and the wrench size required for the bolts.

Multiple Completion Flanges

Figure 4-6 shows two details of a typical 5,000-lb psi-working-pressure segmented flange used for triple or quadruple completions. Table 4-16 lists the cap screw requirements, the API ring gasket number, and the hex wrench size for the cap screws.

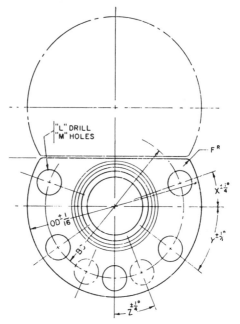

Figure 4-5. 5,000-psi maximum working pressure API segmented flange for dual completions. (Courtesy of American Petroleum Institute.)

Table 4-15
5,000 psi Maximum Working Pressure Segmented
Flanges for Dual Completions

Nominal Size (in.)	Old Nominal Size (in.)	Number of Bolts	Size and Length of Bolts (in.)	Wrench Size for Bolts (in.)	Ring Number
1³/₈	1¹/₄	5	¹/₂ × 4¹/₂	⁷/₈	RX-201
1¹³/₁₆	1³/₄	5	⁵/₈ × 5³/₄	1¹/₁₆	RX-205
2¹/₁₆	2	5	³/₄ × 6	1¹/₄	RX-20
2⁹/₁₆	2¹/₂	5	1 × 7¹/₄	1⁵/₈	RX-210
3¹/₈	3	5	1 × 7³/₄	1⁵/₈	RX-25
4¹/₁₆	4	6	1¹/₈ × 8¹/₄	1¹³/₁₆	RX-215
4¹/₁₆ × 4¹/₄	4 × 4¹/₄	6	1¹/₈ × 8¹/₄	1¹³/₁₆	RX-215

96

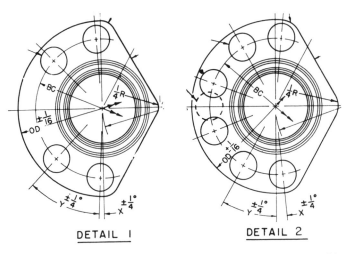

DETAIL 1   DETAIL 2

Figure 4-6. 5,000-psi maximum working pressure API segmented flanges for triple and quadruple completions. (Courtesy of American Petroleum Institute.)

Table 4-16
5,000 psi Maximum Working Pressure Segmented Flanges for Triple and Quadruple Completions

Nominal Size (in.)	Detail Number	Old Nominal Size (in.)	Number of Cap Screws	Size and Length of Cap Screws (in.)	Hex Wrench Size for Cap Screws (in.)	Ring Number
$1^{13}/_{16}$	2	$1^{3}/_{4}$	5	$^{5}/_{8}$-11NC $\times$ $2^{3}/_{4}$	$^{1}/_{2}$	RX-205
$2^{1}/_{16}$	1	2	4	$^{7}/_{8}$-9NC $\times$ $3^{1}/_{4}$	$^{3}/_{4}$	RX-20
$2^{9}/_{16}$	2	$2^{1}/_{2}$	5	1-8UNC $\times$ $3^{1}/_{2}$	$^{3}/_{4}$	RX-210
$3^{1}/_{8}$	2	3	6	$^{7}/_{8}$-9NC $\times$ $3^{1}/_{2}$	$^{3}/_{4}$	RX-25
$4^{1}/_{16}$	2	4	6	1-8UNC $\times$ 4	$^{3}/_{4}$	RX-215
$4^{1}/_{16} \times 4^{1}/_{4}$	2	$4 \times 4^{1}/_{4}$	6	1-8UNC $\times$ 4	$^{3}/_{4}$	RX-215

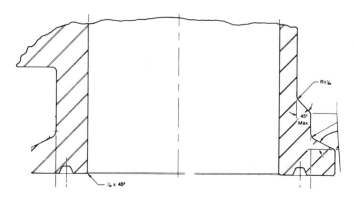

Figure 4-7. Hubs used with API clamp type connectors. (Courtesy of American Petroleum Institute.)

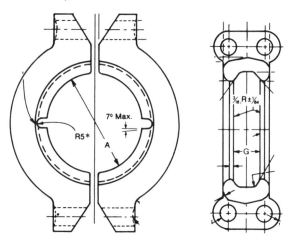

Figure 4-8. API clamp type connectors for use on hubs. (Courtesy of American Petroleum Institute.)

API Hubs and Clamps

Figures 4-7 and 4-8 show hubs and clamp type connectors used by some wellhead manufacturers for 5,000 and 10,000 lb psi pressure ratings.

Hubs are designed by nominal size and bores. Clamps are designated by clamp numbers, 1A through 15A. Clamp numbers 9A through 15A require a spherical washer. The nuts for clamps 1A through 8A have a spherical facing. RX type ring gaskets are used on clamp connectors.

Tables 4-17 and 4-18 show the nominal hub size, hub OD, clamp number, clamp weight, RX gasket number, the bolt size and length, spherical washer size, and the

Table 4-17
API Clamp Type Connectors
Integral Hubs and Clamps
5000 psi Maximum Working Pressure

Nominal Size (in.)	OD of Hub (in.)	Ring Number	Clamp Number	Clamp Weight (lbs)	Nut Size for Clamp (in.)	Wrench Size for Nut (in.)	Washer OD (in.)
2¹/₁₆	5.500	RX-23	1A	35	⁷/₈-9UNC-2B	1⁷/₁₆	—
2⁹/₁₆	6.750	RX-24	2A	45	1-8UNC-2B	1⁵/₈	—
3¹/₈	7.500	RX-27	3A	100	1¹/₈-8UN-2B	1¹³/₁₆	—
4¹/₁₆	9.250	RX-35	4A	56	1¹/₄-8UN-2B	2	—
5¹/₈	11.500	RX-39	5A	67	1³/₈-8UN-2B	2³/₁₆	—
7¹/₁₆	13.625	RX-45	6A	111	1⁵/₈-8UN-2B	2⁹/₁₆	—
9	16.000	RX-49	7A	160	2-8UN-2B	3¹/₈	—
11	18.500	RX-53	8A	242	2¹/₂-8UN-2B	3⁷/₈	—
13³/₈	20.625	RX-57	9A	225	2⁵/₈-8UN-2B	4¹/₁₆	4.06*
16³/₄	25.625	RX-65	10A	345	3¹/₄-8UN-2B	5	5.00

* Clamp sizes 1A through 8A require the use of spherical face nuts.
Clamp sizes 9A and 10A require the use of spherical washers and standard nuts.

Table 4-18
API Clamp Type Connectors
Integral Hubs and Clamps
10,000 psi Maximum Working Pressure

Nominal Size (in.)	OD of Hub (in.)	Ring Number	Clamp Number	Clamp Weight (lbs)	Nut Size for Clamp (in.)	Wrench Size for Nut (in.)	Washer OD (in.)
1 13/16	5.500	RX-20	1A	28	7/8-9UNC-2B	1 7/16	—
2 1/16	6.750	RX-23	2A	35	1-8UNC-2B	1 5/8	—
2 9/16	7.500	RX-24	3A	45	1 1/8-8UN-2B	1 13/16	—
3 1/16	9.250	RX-27	4A	100	1 1/4-8UN-2B	2	—
4 1/16	11.500	RX-35	5A	56	1 3/8-8UN-2B	2 3/16	—
7 1/16	16.000	RX-45	7A	111	2-8UN-2B	3 1/8	—
9	18.500	RX-49	8A	160	2 1/2-8UN-2B	3 7/8	—
11	20.625	RX-53	11A	274	2 3/4-8UN-2B	4 1/4	4.25*
13 5/8	22.468	RX-57	12A	215	3 1/4-8UN-2B	5	5.00
16 3/4	28.000	RX-65	13A	593	3 7/8-8UN-2B	5 15/16	5.94
18 3/4	31.250	RX-69	14A	366	4 1/2-8UN-2B	6 7/8	6.88
21 1/4	34.000	RX-73	15A	850	4 3/4-8UN-2B	7 1/4	7.25

* Clamp sizes 1A through 8A require the use of spherical face nuts.
Clamp sizes 9A and 15A require the use of spherical washers and standard nuts.

Table 4-19
Recommended Bolt Torque for Clamp Type Connectors

Bolt Size	Bolt Tension (lbs)	Makeup Torque (ft lbs)
7/8 -9UNC	16,760	195
1 -8UNC	22,040	292
1 1/8 -8UN	29,120	428
1 1/4 -8UN	37,160	600
1 3/8 -8UN	46,200	815
1 5/8 -8UN	67,200	1382
2 -8UN	106,000	2645
2 1/2 -8UN	171,600	5287
2 5/8 -8UN	190,400	6182
2 3/4 -8UN	210,400	7099
3 1/4 -8UN	299,600	11,685
3 7/8 -8UN	433,200	20,236
4 1/2 -8UN	591,200	32,078
4 3/4 -8UN	660,000	37,745

wrench size required to fit the nut. Table 4-19 is the recommended bolt torque for clamp type connectors.

API Ring Gaskets

API ring gaskets type R, RX, and BX are used for flanges and clamp type connectors. The R and RX gaskets are interchangeable and will fit ANSI flanges where applicable (Standard B-16.20, and API 6A). Uses for the different ring gaskets are shown in Figure 4-9.

Types RX and BX provide a pressure energized seal but are not interchangeable.

It is not recommended to reuse BX150 through BX160 ring gaskets.

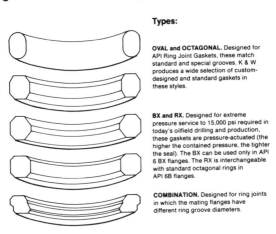

Types:

OVAL and OCTAGONAL. Designed for API Ring Joint Gaskets, these match standard and special grooves. K & W produces a wide selection of custom-designed and standard gaskets in these styles.

BX and RX. Designed for extreme pressure service to 15,000 psi required in today's oilfield drilling and production, these gaskets are pressure-actuated (the higher the contained pressure, the tighter the seal). The BX can be used only in API 6 BX flanges. The RX is interchangeable with standard octagonal rings in API 6B flanges.

COMBINATION. Designed for ring joints in which the mating flanges have different ring groove diameters.

Figure 4-9. Types of API ring gaskets as manufactured by K&W, Inc. (Courtesy of K&W, Inc., a Standco Company.)

Table 4-20
Oval and Octagonal Ring Gasket
Interchange Table for API Type 6B Flanges
with Old Nominal Pipe Sizes

Old Nominal Flange Size	720 960 2,000 3,000 lbs		Old Nominal Flange Size (in.)	5,000 lbs		Old Nominal Flange Size (in.)	2,900 lbs	
	Ring No.			Ring No.			Ring No.	
	R	RX		R	RX		R	RX
1	R-16		1	R-16		1	*R-82	RX-82
1¼	R-18		1¼	R-18		1½	R-84	RX-84
1½	R-20	RX-20	1½	R-20	RX-20	2	R-85	RX-85
2	R-23	RX-23	2	R-24	RX-24	2½	R-86	RX-86
2½	R-26	RX-26	2½	R-27	RX-27	3	R-87	RX-87
3	R-31	RX-31	3	R-35	RX-35	4	R-88	RX-88
3½	R-34	RX-34	3½	R-37	RX-37	3½	R-89	RX-89
4	R-37	RX-37	4	R-39	RX-39	5	R-90	RX-90
5	R-41	RX-41	5	R-44	RX-44	10	R-91	RX-91
6	R-45	RX-45	6	R-46	RX-46			
8	R-49	RX-49	8	R-50	RX-50			
10	R-53	RX-53	10	R-54	RX-54			
12	R-57	RX-57						
14	R-61	RX-61						
16	R-65	RX-65						
18	R-69	RX-69						
20	R-73	RX-73						

*Octagonal is standard in R-80 through R-99.

Gasket Markings

API ring gaskets are marked with the API monogram, the R-number, and type of steel as follows:

Material	Marking
Soft Iron	D (Cadmium plated)
Type 304 SS	S304
Type 316 SS	S316

Table 4-21
BX Gaskets for API Type 6BX Weldneck Flanges

Nominal Flange Bore for 5000 (lbs)	BX-No.	Nominal Flange Bore for 10,000 (lbs)	BX-No.	Nominal Flange Bore for 15,000 (lbs)	BX-No.	Nominal Flange Bore for 20,000 (lbs)	BX-No.
13⅝	BX-160	1 11/16	BX-150	1 11/16	BX-150	1 13/16	BX-151
16¾	BX-162	1 13/16	BX-151	1 13/16	BX-151	2 1/16	BX-152
18¾	BX-163	2 1/16	BX-152	2 1/16	BX-152	2 9/16	BX-153
21¼	BX-165	2 9/16	BX-153	2 9/16	BX-153	3 1/16	BX-154
		3 1/16	BX-154	3 1/16	BX-154	4 1/16	BX-155
		4 1/16	BX-155	4 1/16	BX-155	7 1/16	BX-156
		5⅛	BX-169	7 1/16	BX-156		
		7 1/16	BX-156				
		9	BX-157				
		11	BX-158				
		13⅝	BX-159				
		16¾	BX-162				
		18¾	BX-164				
		21¼	BX-166				

Care of Ring Gaskets

Store ring gaskets on a flat surface, with cardboard or heavy paper as dunnage to separate each gasket for surface protection. Do not hang ring gaskets on pegs or nails. Do not store gaskets on their edges.

Never ship ring gaskets loose or in sacks. Instead, protect the gaskets during shipment by wrapping each one in paper, foil, or in boxes in order to prevent damage to the gasket surface.

5

STAINLESS STEELS

Stainless steels offer a good resistance to certain types of corrosion, and provide acceptable solutions for use in high temperature and sub-zero conditions.

Dimensions and Markings

Fittings made from nickle, aluminum, copper, molybdenum or titanium are the same as the carbon steel fittings described in Chapter 2. However, wall thickness and weights are different for stainless schedules 5S, 10S, 40S, and 80S on sizes of 12 ins. and smaller, which are made in accordance to B36.19. See Table 5-1 for the complete dimensions of stainless steel pipe fittings.

Table 5-1
Dimensions for Stainless Steel Weld Fittings
(Courtesy of Flowline Corp.)

NOM. PIPE SIZE	OUTSIDE DIAMETER (O.D.)	90° ELBOWS Long Radius	90° ELBOWS Short Radius	45° ELBOWS Long Radius		180° RETURNS Long Radius	
		Center to Face (A)	Center to Face (A)	Center to Face (B)	Radius (A)	Center to Center (O)	Back to Face (K)
½	.840	1½	....	⅝	1½	3	1⅞
¾	1.050	1⅛	....	7/16	1⅛	2¼	1 11/16
1	1.315	1½	1	⅞	1½	3	2 3/16
1¼	1.660	1⅞	1¼	1	1⅞	3¾	2¾
1½	1.900	2¼	1½	1⅛	2¼	4½	3¼
2	2.375	3	2	1⅜	3	6	4 3/16
2½	2.875	3¾	2½	1¾	3¾	7½	5 3/16
3	3.500	4½	3	2	4½	9	6¼
3½	4.000	5¼	3½	2¼	5¼	10½	7¼
4	4.500	6	4	2½	6	12	8¼
5	5.563	7½	5	3⅛	7½	15	10 5/16
6	6.625	9	6	3¾	9	18	12 5/16
8	8.625	12	8	5	12	24	16 5/16
10	10.750	15	10	6¼	15	30	20⅜
12	12.750	18	12	7½	18	36	24⅜
14	14.000	21	14	8¾	21	42	28
16	16.000	24	16	10	24	48	32
18	18.000	27	18	11¼	27	54	36
20	20.000	30	20	12½	30	60	40
24	24.000	36	24	15	36	72	48

NOM. PIPE SIZE	OUTSIDE DIAMETER (O.D.)	STRAIGHT TEES	CAPS		STUB ENDS			STRAIGHT CROSSES
		Center to End (C)	Length (E)	Length (E-1)†	Lap Diameter (G)	(Long) Length (F)	(Short) Length (F)	Center to End (C)
½	.840	1	1	....	1⅜	3	2	....
¾	1.050	1⅛	1	....	1 11/16	3	2	....
1	1.315	1½	1½	1½	2	4	2	1½
1¼	1.660	1⅞	1½	1½	2½	4	2	1⅞
1½	1.900	2¼	1½	1½	2⅞	4	2	2¼
2	2.375	2½	1½	1¾	3⅝	6	2½	2½
2½	2.875	3	1½	2	4⅛	6	2½	3
3	3.500	3⅜	2	2½	5	6	2½	3⅜
3½	4.000	3¾	2½	3	5½	6	3	3¾
4	4.500	4⅛	2½	3	6 3/16	6	3	4⅛
5	5.563	4⅞	3	3½	7 5/16	8	3	4⅞
6	6.625	5⅝	3½	4	8½	8	3½	5⅝
8	8.625	7	4	5	10⅝	8	4	7
10	10.750	8½	5	6	12¾	10	5	8½
12	12.750	10	6	7	15	10	6	10
14	14.000	11	6½	7½	16¼	12	6	11
16	16.000	12	7	8	18½	12	6	12
18	18.000	13½	8	9	21	12	6	13½
20	20.000	15	9	10	23	12	6	15
24	24.000	17	10½	12	27¼	12	6	17

†Use length E-1 where wall thickness is greater than for Schedule 80S.

(Continued on next page)

Table 5-1 continued

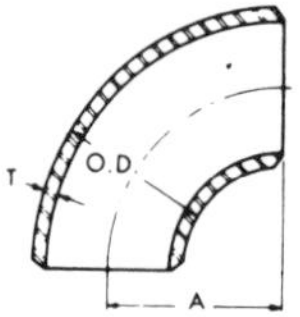

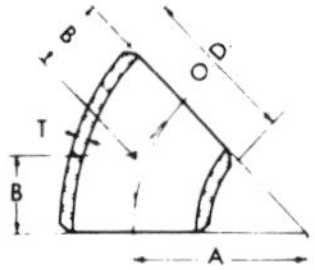

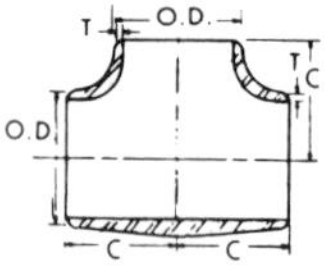

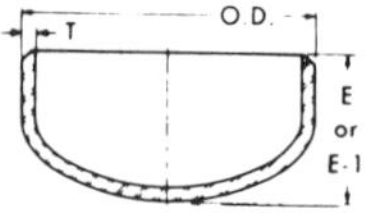

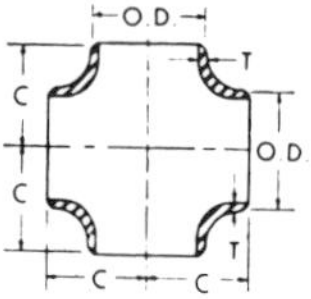

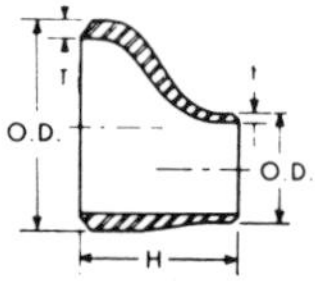

NOM. PIPE SIZE	REDUCERS Concentric & Eccentric Length (H)	REDUCING OUTLET TEES * Center to End of Run (C)	Center to End of Outlet (M)
½ x ¼	1½	1	1
x ⅜	1½	1	1
¾ x ⅜	2	1⅛	1⅛
x ½	2	1⅛	1⅛
1 x ⅜	2	1½	1½
x ½	2	1½	1½
x ¾	2	1½	1½
1¼ x ½	2	1⅞	1⅞
x ¾	2	1⅞	1⅞
x 1	2	1⅞	1⅞
1½ x ½	2½	2¼	2¼
x ¾	2½	2¼	2¼
x 1	2½	2¼	2¼
x 1¼	2½	2¼	2¼
2 x ¾	3	2½	1¾
x 1	3	2½	2
x 1¼	3	2½	2¼
x 1½	3	2½	2⅜
2½ x 1	3½	3	2¼
x 1¼	3½	3	2½
x 1½	3½	3	2⅝
x 2	3½	3	2¾
3 x 1	3½	3⅜	2⅝
x 1½	3½	3⅜	2⅞
x 2	3½	3⅜	3
x 2½	3½	3⅜	3¼

NOM. PIPE SIZE	REDUCERS Concentric & Eccentric Length (H)	REDUCING OUTLET TEES * Center to End of Run (C)	Center to End of Outlet (M)
3½ x 1¼	4	....	....
x 1½	4	3¾	3⅛
x 2	4	3¾	3¼
x 2½	4	3¾	3½
x 3	4	3¾	3⅝
4 x 1½	4	4⅛	3⅜
x 2	4	4⅛	3½
x 2½	4	4⅛	3¾
x 3	4	4⅛	3⅞
x 3½	4	4⅛	4
5 x 2	5	4⅞	4⅛
x 2½	5	4⅞	4¼
x 3	5	4⅞	4⅜
x 3½	5	4⅞	4½
x 4	5	4⅞	4⅝

(Continued on next page)

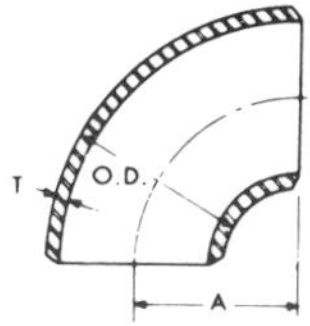

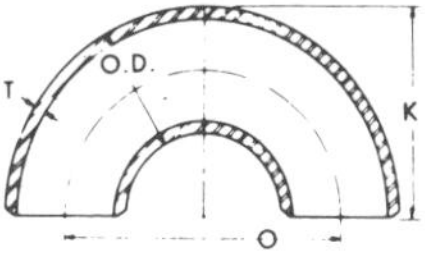

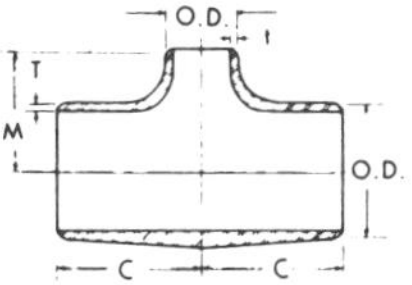

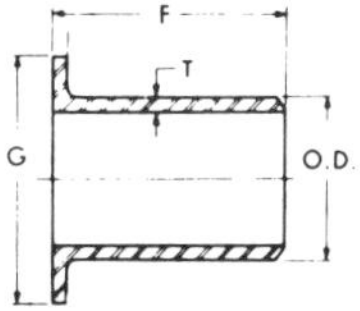

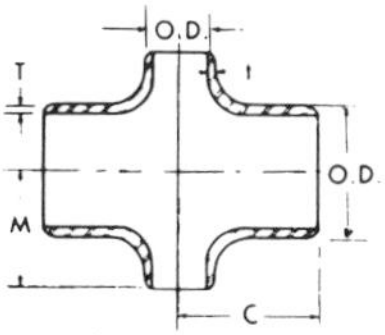

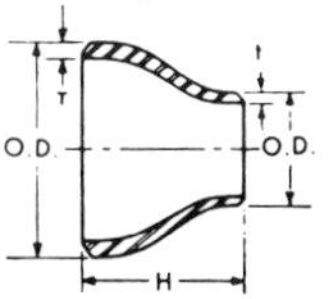

Table 5-1 continued

NOM. PIPE SIZE	REDUCERS	REDUCING OUTLET TEES *	
	Concentric & Eccentric Length (H)	Center to End of Run (C)	Center to End of Outlet (M)
6 x 2½	5½	5⅝	4¾
x 3	5½	5⅝	4⅞
x 3½	5½	5⅝	5
x 4	5½	5⅝	5⅛
x 5	5½	5⅝	5⅜
8 x 3	6	7	6
x 3½	6	7	6
x 4	6	7	6⅛
x 5	6	7	6⅜
x 6	6	7	6⅝
10 x 4	7	8½	7¼
x 5	7	8½	7½
x 6	7	8½	7⅝
x 8	7	8½	8

NOM. PIPE SIZE	REDUCERS	REDUCING OUTLET TEES *	
	Concentric & Eccentric Length (H)	Center to End of Run (C)	Center to End of Outlet (M)
12 x 5	8	10	8½
x 6	8	10	8⅝
x 8	8	10	9
x 10	8	10	9½
14 x 6	13	11	9⅜
x 8	13	11	9¾
x 10	13	11	10⅛
x 12	13	11	10⅝
16 x 6	14	12	10⅜
x 8	14	12	10¾
x 10	14	12	11⅛
x 12	14	12	11⅝
x 14	14	12	12
18 x 8	15	13½	11¾
x 10	15	13½	12⅛
x 12	15	13½	12⅝
x 14	15	13½	13
x 16	15	13½	13
20 x 8	20	15	12¾
x 10	20	15	13⅛
x 12	20	15	13⅝
x 14	20	15	14
x 16	20	15	14
x 18	20	15	14½
24 x 10	20	17	15⅛
x 12	20	17	15⅝
x 14	20	17	16
x 16	20	17	16
x 18	20	17	16½
x 20	20	17	17

*Use same dimensions for reducing outlet crosses.

All dimensions are in inches and conform to ASA B16.9 and MSS SP-43, where applicable.

Figure 5-1. Standard markings for stainless steel fittings. (Courtesy of Flowline Corp.)

Markings of Fittings

Figure 5-1 illustrates a stainless steel 90-degree weld elbow long radius with standard markings:

7071	(heat or batch number)
4″ Sch. 40S	(size and pipe schedule)
Flowline	(trademark of the manufacturer)
.237″ Wall	(wall thickness of S/40S pipe)
WP304L	(ASTM specification design)

Types of Stainless Steel

There are over forty types of stainless steels. Three basic types account for half of the stainless steel used. These are the 300 and 400 series of stainless.
108

The most likely types of stainless steel used in ANSI systems will be 304, 304L, 316, and 316L. The "L" designation is for a low-carbon content in the steel.

Screwed fittings are usually 304 or 316 stainless. Valves are usually 316 stainless. Pipe and flanges may be any of the above types. Stainless steel stud bolts are not common.

Carbon steel lap joint flanges are used with stainless steel stub ends when feasible to limit the use of the higher priced stainless flanges. See Table 5-3 for dimensions of stub ends.

Stainless Steel Fittings

Figures 5-2 through 5-5 show the most commonly used weld fittings.

Identification markings are identical to the requirements for carbon steel flanges, i.e., trademark, type of material, size, pressure rating, and bore. Descriptions are the same as for carbon steel flanges.

Storage of Stainless Flanges

The same methods described in Chapter 2 for storing or shipping carbon steel flanges applies as well for storing stainless steel flanges. However, it is preferable to store stainless flanges indoors if possible to protect the flange face and gasket surface. It is not necessary to lubricate the face of any stainless steel flange.

(Text continued on page 110)

LONG RADIUS

SHORT RADIUS

90° ELBOWS

90° ELBOWS

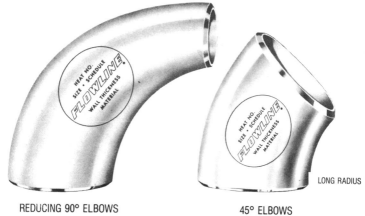

LONG RADIUS

REDUCING 90° ELBOWS

45° ELBOWS

Figure 5-2. Stainless steel elbows. (Courtesy of Flowline Corp.)

110

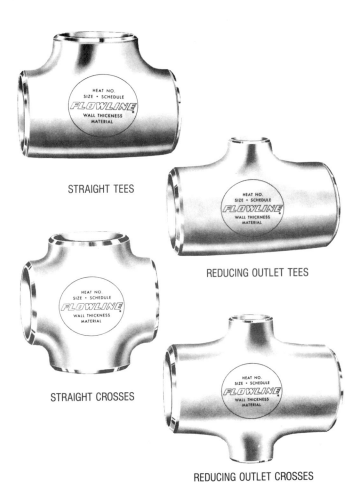

STRAIGHT TEES

HEAT NO.
SIZE • SCHEDULE
FLOWLINE
WALL THICKNESS
MATERIAL

REDUCING OUTLET TEES

HEAT NO.
SIZE • SCHEDULE
FLOWLINE
WALL THICKNESS
MATERIAL

STRAIGHT CROSSES

HEAT NO.
SIZE • SCHEDULE
FLOWLINE
WALL THICKNESS
MATERIAL

REDUCING OUTLET CROSSES

Figure 5-3. Stainless steel tees and crosses. (Courtesy of Flowline Corp.)

111

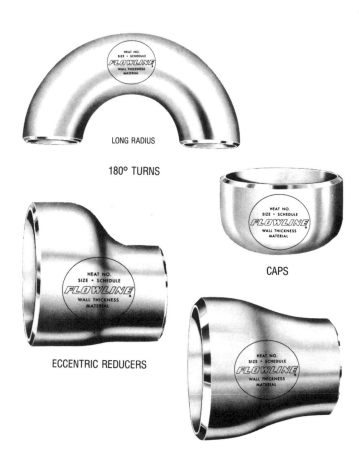

LONG RADIUS

180° TURNS

CAPS

ECCENTRIC REDUCERS

CONCENTRIC REDUCERS

Figure 5-4. Stainless steel returns, caps, and reducers. (Courtesy of Flowline Corp.)

112

FLANGES

STUB ENDS—MSS SHORT LENGTHS

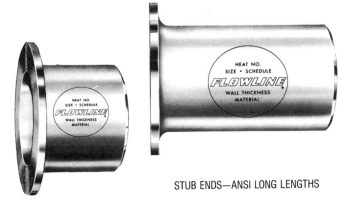

STUB ENDS—ANSI LONG LENGTHS

TYPE C STUB ENDS

Figure 5-5. Stainless steel stub ends and flange. (Courtesy of Flowline Corp.)

Stainless Steel Pipe

The common schedules of stainless steel pipe are 5S, 10S, 40S, and 80S. Table 5-2 is a complete chart of stainless steel pipe schedules and other important data. Stainless pipe is commonly referred to and described by size, schedule, and the wall thickness. Weight per foot is seldom mentioned.

Example: 8-in. S/10S .109″ WT.
8-in. S/40S .322″ WT.
1-in. S/80S .179″ WT.

Care of Stainless Steel Pipe

Extra care of stainless steel pipe during the entire cycle of receipt, storage, fabrication, and shipment is a requirement that cannot be over emphasized. Stainless pipe has a finished surface and thin walls. Pipe should be stored indoors, if possible, and on carpet-covered pipe racks.

The pipe should be stored by type, size and schedule. Suitable dunnage should be used between each layer of pipes.

Nylon slings, not chains or cable slings, should be used to handle the pipe. A special protective harness is available for use when shipping stainless pipe for protection from damage, which can be caused by chains and binders.

Stainless pipe should never be dropped or bumped against other joints. *(Text continued on page 118)*

114

Table 5-2
Dimensions for Stainless Steel Pipe
(Courtesy of Flowline Corp.)

Nominal Pipe Size	Outside Diameter	SCHEDULE 5S		SCHEDULE 10S		SCHEDULE 10	
		Wall Thick.	Inside Diam.	Wall Thick.	Inside Diam.	Wall Thick.	Inside Diam.
1/8	.405	....	....	.049	.307	....	....
1/4	.540	....	....	.065	.410	....	....
3/8	.675	....	....	.065	.545	....	....
1/2	.840	.065	.710	.083	.674	....	....
3/4	1.050	.065	.920	.083	.884	....	....
1	1.315	.065	1.185	.109	1.097	....	....
1 1/4	1.660	.065	1.530	.109	1.442	....	....
1 1/2	1.900	.065	1.770	.109	1.682	....	....
2	2.375	.065	2.245	.109	2.157	....	....
2 1/2	2.875	.083	2.709	.120	2.635	....	....
3	3.500	.083	3.334	.120	3.260	....	....
3 1/2	4.000	.083	3.834	.120	3.760	....	....
4	4.500	.083	4.334	.120	4.260	....	....
5	5.563	.109	5.345	.134	5.295	....	....
6	6.625	.109	6.407	.134	6.357	....	....
8	8.625	.109	8.407	.148	8.329	....	....
10	10.750	.134	10.482	.165	10.420	....	....
12	12.750	.156	12.438	.180	12.390	....	....
14	14.000	.156(A)	13.688	.188(A)*	13.624	.250	13.500
16	16.000	.165(A)	15.670	.188(A)	15.624	.250	15.500
18	18.000	.165(A)	17.670	.188(A)	17.624	.250	17.500
20	20.000	.188(A)	19.624	.218(A)	19.564	.250	19.500
24	24.000	.218(A)	23.564	.250(A)	23.500	.250	23.500

Nominal Pipe Size	Outside Diameter	SCHEDULE 20		SCHEDULE 30		SCHEDULE 40S AND STANDARD WT. (B)	
		Wall Thick.	Inside Diam.	Wall Thick.	Inside Diam.	Wall Thick.	Inside Diam.
1/8	.405	....	....	....	....	.068	.269
1/4	.540	....	....	....	....	.088	.364
3/8	.675	....	....	....	....	.091	.493
1/2	.840	....	....	....	....	.109	.622
3/4	1.050	....	....	....	....	.113	.824
1	1.315	....	....	....	....	.133	1.049
1 1/4	1.660	....	....	....	....	.140	1.380
1 1/2	1.900	....	....	....	....	.145	1.610
2	2.375	....	....	....	....	.154	2.067
2 1/2	2.875	....	....	....	....	.203	2.469
3	3.500	....	....	....	....	.216	3.068
3 1/2	4.000	....	....	....	....	.226	3.548
4	4.500	....	....	....	....	.237	4.026
5	5.563	....	....	....	....	.258	5.047
6	6.625	....	....	....	....	.280	6.065
8	8.625	.250	8.125	.277	8.071	.322	7.981
10	10.750	.250	10.250	.307	10.136	.365	10.020
12	12.750	.250	12.250	.330	12.090	.375	12.000
14	14.000	.312	13.376	.375	13.250	.375(1)	13.250
16	16.000	.312	15.376	.375	15.250	.375(1)	15.250
18	18.000	.312	17.376	.438	17.124	.375(1)	17.250
20	20.000	.375	19.250	.500	19.000	.375(1)	19.250
24	24.000	.375	23.250	.562	22.876	.375(1)	23.250

(Continued on next page)

Table 5-2 continued

Nominal Pipe Size	Outside Diameter	SCHEDULE 40(B)		SCHEDULE 60		SCHEDULE 80S AND EXTRA STRONG (C)	
		Wall Thick.	Inside Diam.	Wall Thick.	Inside Diam.	Wall Thick.	Inside Diam.
⅛	.405	.068	.269	....	....	.095	.215
¼	.540	.088	.364	....	....	.119	.302
⅜	.675	.091	.493	....	....	.126	.423
½	.840	.109	.622	....	....	.147	.546
¾	1.050	.113	.824	....	....	.154	.742
1	1.315	.133	1.049	....	....	.179	.957
1¼	1.660	.140	1.380	....	....	.191	1.278
1½	1.900	.145	1.610	....	....	.200	1.500
2	2.375	.154	2.067	....	....	.218	1.939
2½	2.875	.203	2.469	....	....	.276	2.323
3	3.500	.216	3.068	....	....	.300	2.900
3½	4.000	.226	3.548	....	....	.318	3.364
4	4.500	.237	4.026	....	....	.337	3.826
5	5.563	.258	5.047	....	....	.375	4.813
6	6.625	.280	6.065	....	....	.432	5.761
8	8.625	.322	7.981	.406	7.813	.500	7.625
10	10.750	.365	10.020	.500	9.750	.500	9.750
12	12.750	.406	11.938	.562	11.626	.500	11.750
14	14.000	.438	13.124	.593	12.814	.500(2)	13.000
16	16.000	.500	15.000	.656	14.688	.500(2)	15.000
18	18.000	.562	16.876	.750	16.500	.500(2)	17.000
20	20.000	.593	18.814	.812	18.376	.500(2)	19.000
24	24.000	.687	22.626	.968	22.064	.500(2)	23.000

Nominal Pipe Size	Outside Diameter	SCHEDULE 80 (C)		SCHEDULE 100		SCHEDULE 120	
		Wall Thick.	Inside Diam.	Wall Thick.	Inside Diam.	Wall Thick.	Inside Diam.
⅛	.405	.095	.215	....	....	....	....
¼	.540	.119	.302	....	....	....	....
⅜	.675	.126	.423	....	....	....	....
½	.840	.147	.546	....	....	....	....
¾	1.050	.154	.742	....	....	....	....
1	1.315	.179	.957	....	....	....	....
1¼	1.660	.191	1.278	....	....	....	....
1½	1.900	.200	1.500	....	....	....	....
2	2.375	.218	1.939	....	....	....	....
2½	2.875	.276	2.323	....	....	....	....
3	3.500	.300	2.900	....	....	....	....
3½	4.000	.318	3.364	....	....	....	....
4	4.500	.337	3.826	....	....	.438	3.624
5	5.563	.375	4.813	....	....	.500	4.563
6	6.625	.432	5.761	....	....	.562	5.501
8	8.625	.500	7.625	.593	7.439	.718	7.189
10	10.750	.593	9.564	.718	9.314	.843	9.064
12	12.750	.687	11.376	.843	11.064	1.000	10.750
14	14.000	.750	12.500	.937	12.126	1.093	11.814
16	16.000	.843	14.314	1.031	13.938	1.218	13.564
18	18.000	.937	16.126	1.156	15.688	1.375	15.250
20	20.000	1.031	17.938	1.281	17.438	1.500	17.000
24	24.000	1.218	21.564	1.531	20.938	1.812	20.376

(Continued on next page)

Table 5-2 continued

Nominal Pipe Size	Outside Diameter	SCHEDULE 140		SCHEDULE 160		DOUBLE X STRONG	
		Wall Thick.	Inside Diam.	Wall Thick.	Inside Diam.	Wall Thick.	Inside Diam.
1/8	.405	....	....	....	....	....	....
1/4	.540	....	....	....	....	....	....
3/8	.675	....	....	....	....	....	....
1/2	.840	....	....	.187	.466	.294	.252
3/4	1.050	....	....	.218	.614	.308	.434
1	1.315	....	....	.250	.815	.358	.599
1 1/4	1.660	....	....	.250	1.160	.382	.896
1 1/2	1.900	....	....	.281	1.338	.400	1.100
2	2.375	....	....	.343	1.689	.436	1.503
2 1/2	2.875	....	....	.375	2.125	.552	1.771
3	3.500	....	....	.438	2.624	.600	2.300
3 1/2	4.000	....	....	....	....	....	....
4	4.500	....	....	.531	3.438	.674	3.152
5	5.563	....	....	.625	4.313	.750	4.063
6	6.625	....	....	.718	5.189	.864	4.897
8	8.625	.812	7.001	.906	6.813	.875	6.875
10	10.750	1.000	8.750	1.125	8.500	....	....
12	12.750	1.125	10.500	1.312	10.126	....	....
14	14.000	1.250	11.500	1.406	11.188	....	....
16	16.000	1.438	13.124	1.593	12.814	....	....
18	18.000	1.562	14.876	1.781	14.438	....	....
20	20.000	1.750	16.500	1.968	16.064	....	....
24	24.000	2.062	19.876	2.343	19.314	....	....

All dimensions are in inches.

Dimensions for Standard Weight, Extra Strong, Double Extra Strong, Schedules 10, 20, 30, 40, 60, 80, 100, 120, 140 and 160 are in conformance with A.S.A. B36.10.

Dimensions for Schedules 5S, 10S, 40S, and 80S are in conformance with A.S.A. B36.19.

(A) Proposed wall thickness for Schedules 5S and 10S.

(B) Wall thicknesses for Schedules 40, 40S, and Standard Weight are identical through 10" size.

(C) Wall thicknesses for Schedules 80, 80S, and Extra Strong are identical through 8" size.

(1) Thickness agrees with that for Standard Weight Pipe (A.S.A. B36.10); not included in Schedule 40S.

(2) Thickness agrees with that for Extra Strong Pipe (A.S.A. B36.10); not included in Schedule 80S.

Table 5-3
Dimensions of Stainless Steel Stub Ends
(Courtesy of Flowline Corp.)

NOM. PIPE SIZE	OUTSIDE DIAMETER (O.D.)	LAP DIAMETER (G)	LENGTH (F)		RADIUS		SCHEDULE 5S Featherweight		
			Long	Short	A	B	THICKNESS		Stainless Steel Approx. Wt. in Pounds* Short Length
							WALL (T)	LAP (t)	
½	.840	1⅜	3	2	⅛	¹⁄₃₂	.065	.084	.12
¾	1.050	1¹¹⁄₁₆	3	2	⅛	¹⁄₃₂	.065	.086	.14
1	1.315	2	4	2	⅛	¹⁄₃₂	.065	.093	.18
1¼	1.660	2½	4	2	³⁄₁₆	¹⁄₃₂	.065	.095	.28
1½	1.900	2⅞	4	2	¼	¹⁄₃₂	.065	.097	.33
2	2.375	3⅝	6	2½	⁵⁄₁₆	¹⁄₃₂	.065	.100	.49
2½	2.875	4⅛	6	2½	⁵⁄₁₆	¹⁄₃₂	.083	.130	.67
3	3.500	5	6	2½	⅜	¹⁄₃₂	.083	.134	.91
3½	4.000	5½	6	3	⅜	¹⁄₃₂	.083	.137	1.18
4	4.500	6³⁄₁₆	6	3	⁷⁄₁₆	¹⁄₃₂	.083	.140	1.37
5	5.563	7⁵⁄₁₆	8	3	⁷⁄₁₆	¹⁄₁₆	.109	.168	1.89
6	6.625	8½	8	3½	½	¹⁄₁₆	.109	.175	3.45
8	8.625	10⅝	8	4	½	¹⁄₁₆	.109	.187	5.34
10	10.750	12¾	10	5	½	¹⁄₁₆	.134	.221	8.35
12	12.750	15	10	6	½	¹⁄₁₆	.156	.249	13.34
14	14.000	16¼	12	6	½	¹⁄₁₆	.156⁽¹⁾	.249	14.00
16	16.000	18½	12	6	½	¹⁄₁₆	.165⁽¹⁾	.249	17.50
18	18.000	21	12	6	½	¹⁄₁₆	.165⁽¹⁾	.249	25.25
20	20.000	23	12	6	½	¹⁄₁₆	.188⁽¹⁾	.249	28.00
22	22.000	25¼	12	6	½	¹⁄₁₆	.188⁽¹⁾	.249	30.75
24	24.000	27¼	12	6	½	¹⁄₁₆	.218⁽¹⁾	.249	34.50

(Continued on next page)

FLOWLINE TYPE "A" STUB ENDS

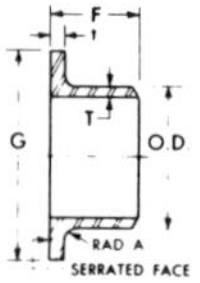

MADE IN CONFORMANCE WITH A.S.A. B16.9 AND M.S.S. SP.-43 WHERE APPLICABLE EXCEPT THICKNESS OF CERTAIN LAPS HAS BEEN INCREASED

Table 5-3 continued

NOM. PIPE SIZE	OUTSIDE DIAMETER (O.D.)	LAP DIAMETER (G)	LENGTH (F) Long	LENGTH (F) Short	RADIUS A	RADIUS B	SCHEDULE 10S Light I.P.S. THICKNESS WALL (T)	SCHEDULE 10S Light I.P.S. THICKNESS LAP (t)	Stainless Steel Approx. Wt. in Pounds* Short Length
½	.840	1 ⅜	3	2	⅛	¹⁄₃₂	.083	.095	.16
¾	1.050	1¹¹⁄₁₆	3	2	⅛	¹⁄₃₂	.083	.097	.18
1	1.315	2	4	2	⅛	¹⁄₃₂	.109	.120	.30
1 ¼	1.660	2 ½	4	2	³⁄₁₆	¹⁄₃₂	.109	.124	.48
1 ½	1.900	2 ⅞	4	2	¼	¹⁄₃₂	.109	.126	.55
2	2.375	3 ⅝	6	2 ½	⁵⁄₁₆	¹⁄₃₂	.109	.130	.80
2 ½	2.875	4 ⅛	6	2 ½	⁵⁄₁₆	¹⁄₃₂	.120	.156	.96
3	3.500	5	6	2 ½	⅜	¹⁄₃₂	.120	.161	1.34
3 ½	4.000	5 ½	6	3	⅜	¹⁄₃₂	.120	.165	1.72
4	4.500	6³⁄₁₆	6	3	⁷⁄₁₆	¹⁄₃₂	.120	.169	1.99
5	5.563	7⁵⁄₁₆	8	3	⁷⁄₁₆	¹⁄₁₆	.134	.186	2.26
6	6.625	8 ½	8	3 ½	½	¹⁄₁₆	.134	.194	4.25
8	8.625	10 ⅝	8	4	½	¹⁄₁₆	.148	.218	6.73
10	10.750	12 ¾	10	5	½	¹⁄₁₆	.165	.245	10.31
12	12.750	15	10	6	½	¹⁄₁₆	.180	.260	14.39
14	14.000	16 ¼	12	6	½	¹⁄₁₆	.188[1]	.260	16.75
16	16.000	18 ½	12	6	½	¹⁄₁₆	.188[1]	.260	20.00
18	18.000	21	12	6	½	¹⁄₁₆	.188[1]	.260	21.50
20	20.000	23	12	6	½	¹⁄₁₆	.218[1]	.260	28.25
22	22.000	25 ¼	12	6	½	¹⁄₁₆	.218[1]	.260	31.00
24	24.000	27 ¼	12	6	½	¹⁄₁₆	.250[1]	.260	39.75

(Continued on next page)

FLOWLINE TYPE "B" STUB ENDS

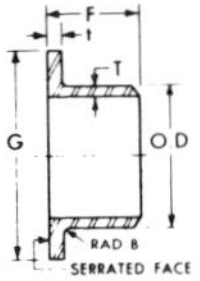

MADE IN
CONFORMANCE
WITH A.S.A. B16.9
AND M.S.S. SP.-43
WHERE APPLICABLE
EXCEPT THICKNESS
OF CERTAIN LAPS
HAS BEEN INCREASED

119

Table 5-3 continued

NOM. PIPE SIZE	OUTSIDE DIAMETER (O.D.)	LAP DIAMETER (G)	LENGTH (F) Long	LENGTH (F) Short	RADIUS A	RADIUS B	SCHEDULE 40S Standard I.P.S. THICKNESS WALL (T)	THICKNESS LAP (t)	Stainless Steel Approx. Wt. in Pounds* Long Length	Stainless Steel Approx. Wt. in Pounds* Short Length	Alum Appro in Pe Long
½	.840	1⅜	3	2	⅛	1/32	.109	.109	.32	.24	
¾	1.050	1¹¹/₁₆	3	2	⅛	1/32	.113	.113	.45	.36	
1	1.315	2	4	2	⅛	1/32	.133	.133	.65	.37	
1¼	1.660	2½	4	2	3/16	1/32	.140	.140	1.00	.62	
1½	1.900	2⅞	4	2	¼	1/32	.145	.145	1.20	.75	
2	2.375	3⅝	6	2½	5/16	1/32	.154	.154	2.25	1.29	
2½	2.875	4⅛	6	2½	5/16	1/32	.203	.203	3.41	1.71	1
3	3.500	5	6	2½	⅜	1/32	.216	.216	4.67	2.46	1
3½	4.000	5½	6	3	⅜	1/32	.226	.226	5.58	3.30	1.
4	4.500	6³/₁₆	6	3	7/16	1/32	.237	.237	6.70	4.06	2
5	5.563	7⁵/₁₆	8	3	7/16	1/16	.258	.258	10.75	4.64	3
6	6.625	8½	8	3½	½	1/16	.280	.280	16.18	9.06	5.
8	8.625	10⅝	8	4	½	1/16	.322	.322	25.50	15.98	8.
10	10.750	12¾	10	5	½	1/16	.365	.365	40.00	23.13	14.
12	12.750	15	10	6	½	1/16	.375	.375	47.00	30.48	16.
14	14.000	16¼	12	6	½	1/16	.375(2)	.375	60.00	35.00	21.
16	16.000	18½	12	6	½	1/16	.375(2)	.375	69.75	41.75	24.
18	18.000	21	12	6	½	1/16	.375(2)	.375	80.50	48.00	28.
20	20.000	23	12	6	½	1/16	.375(2)	.375	91.25	54.50	32.
22	22.000	25¼	12	6	½	1/16	.375(2)	.375	99.75	59.50	35.
24	24.000	27¼	12	6	½	1/16	.375(2)	.375	113.25	67.50	39.

TYPE "C" STUB ENDS

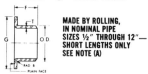

MADE BY ROLLING,
IN NOMINAL PIPE
SIZES ½" THROUGH 12"—
SHORT LENGTHS ONLY
SEE NOTE (A)

Table 5-3 continued

NOM. PIPE SIZE	OUTSIDE DIAMETER (O.D.)	LAP DIAMETER (G)	LENGTH (F)		RADIUS		SCHEDULE 80S Extra Heavy I.P.S.			
							THICKNESS		Stainless Steel Approx. Wt. in Pounds* Long Length	Aluminum Approx. Wt. in Pounds Long Length
			Long	Short	A	B	WALL (T)	LAP (t)		
½	.840	1 ⅜	3	2	⅛	¹⁄₃₂	.147	.187	.38	.13
¾	1.050	1¹¹⁄₁₆	3	2	⅛	¹⁄₃₂	.154	.187	.51	.18
1	1.315	2	4	2	⅛	¹⁄₃₂	.179	.187	.87	.31
1 ¼	1.660	2 ½	4	2	³⁄₁₆	¹⁄₃₂	.191	.191	1.35	.48
1 ½	1.900	2 ⅞	4	2	¼	¹⁄₃₂	.200	.200	1.54	.54
2	2.375	3 ⅝	6	2 ½	⁵⁄₁₆	¹⁄₃₂	.218	.218	3.10	1.09
2 ½	2.875	4 ⅛	6	2 ½	⁵⁄₁₆	¹⁄₃₂	.276	.276	4.64	1.64
3	3.500	5	6	2 ½	⅜	¹⁄₃₂	.300	.300	6.36	2.25
3 ½	4.000	5 ½	6	3	⅜	¹⁄₃₂	.318	.318	7.70	2.72
4	4.500	6³⁄₁₆	6	3	⁷⁄₁₆	¹⁄₃₂	.337	.337	9.37	3.31
5	5.563	7⁵⁄₁₆	8	3	⁷⁄₁₆	¹⁄₁₆	.375	.375	16.50	5.82
6	6.625	8 ½	8	3 ½	½	¹⁄₁₆	.432	.432	22.56	7.96
8	8.625	10 ⅝	8	4	½	¹⁄₁₆	.500	.500	34.50	12.18
10	10.750	12 ¾	10	5	½	¹⁄₁₆	.500	.500	54.00	19.06
12	12.750	15	10	6	½	¹⁄₁₆	.500	.500	64.50	22.77
14	14.000	16 ¼	12	6	½	¹⁄₁₆	.500(3)	.500	82.00	28.75
16	16.000	18 ½	12	6	½	¹⁄₁₆	.500(3)	.500	96.50	33.80
18	18.000	21	12	6	½	¹⁄₁₆	.500(3)	.500	108.50	38.00
20	20.000	23	12	6	½	¹⁄₁₆	.500(3)	.500	119.50	42.00
22	22.000	25 ¼	12	6	½	¹⁄₁₆	.500(3)	.500	128.25	45.00
24	24.000	27 ¼	12	6	½	¹⁄₁₆	.500(3)	.500	148.00	52.00

Cut-offs at fabrications shops should be re-marked if necessary with the standard markings, and then returned to stock for future issue.

Marking Stainless Pipe

Stainless steel pipe and fittings should never be painted because they do not require a protective coating. In addition, identification would be more difficult.

Be sure all stainless steel items are marked with good identification code numbers. Leave any vendor markings on stainless pipe for future identification purposes. Mark your company codes with stencil ink that includes the type, size, schedule, and purchase order item number on each joint. (The stencil ink is available in aerosol cans). You may also use good quality bar code markings. Place codes inside each end of each joint of pipe.

To further ensure the correct identification of stainless steels, a color code scheme is used. A suggested color code appears in Table 5-4. Stencil inks are used to apply the color code in designs of your choice such as lines, dots, triangles, etc.

Identification Tests

If a stainless item is not identified for some reason, besides a laboratory test, craftsmen can perform certain tests at jobsites.

122

Table 5-4
Color Code Chart

Type of steel	Stencil Ink Colors
−21 degree F to −50 degree F Carbon Steel	Yellow
Special Carbon Steel (Project stated as special)	Red
3½% Nickle	White
Type 304 Stainless Steel	Red
Type 304L Stainless Steel	Yellow
Type 316 Stainless Steel	Blue
Type 316L Stainless Steel	Green
Type 309 Stainless Steel	Red & Yellow
Type 310 Stainless Steel	Red & Blue
Carpenter 20 Stainless Steel	Black
Carbon-½% Molybdenum	Red & White
1% Chrome-½% Molybdenum	Yellow & White
1¼% Chrome-½% Molybdenum	Green & White
2¼% Chrome-1% Molybdenum	Green & Yellow
5% Chrome-½% Molybdenum	Orange
9% Chrome-1% Molybdenum	Orange & Yellow
AF-22-65 Duplex	Blue & Yellow
AF-22-100 Duplex	Red & Yellow
AF-22-130 Duplex	Black & Yellow
MW CR-13	Blue & Orange

Magnet Tests

The 300 series of stainless steels are non-magnetic in most cases to a hand magnet. The 400 series of stainless steels are magnetic.

Chemical Tests

Saturated copper sulphate solution deposits metallic copper on non-stainless steel in about five minutes. On stainless steels there will not be a deposit. (To apply the copper sulphate solution for the test, clean a small area of the steel with emory cloth, then apply a few drops of the solution to the abraded area.)

Other Tests

How to distinguish 302 and 304 from 316 and 317 stainless steels is described in Table 5-5 along with other types of tests.

Figure 5-6 illustrates a tool called the WT Alloy Separator.® When the probe is touched to any metal, the probe creates an instant thermocouple voltage that is unique for each metal containing sufficient differences in chemistry or crystalline structure. It is manufactured by Technicorp of Wayne, N.J.

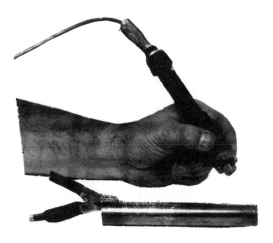

Figure 5-6. Electronic metal tester. (Courtesy of Technicorp-Wayne, New Jersey.)
124

Table 5-5
Identification Tests for Stainless Steels
(Courtesy of Uddeholm Steel Corp.)

	AISI TYPE	GROUP	MAGNET TEST	SPARK TEST	HARDNESS TEST
CHROMIUM-NICKEL GRADES	302	Austenitic	NON-MAGNETIC	Short, reddish, with few forks	Under 165 Brinell after heating to 1800°F and water quench
	303	Austenitic		Short, reddish, with few forks	
	303Se	Austenitic		Short, reddish, with few forks	
	304	Austenitic		Short, reddish, with few forks	
	308	Austenitic		Full red without many forks	
	309	Austenitic		Full red without many forks	
	310	Austenitic		Full red without many forks	
	316	Austenitic		Short, reddish, with few forks	
	317	Austenitic			
	321	Austenitic			
	347	Austenitic			
CHROMIUM GRAD	410	Martensitic	MAGNETIC	Long white with few forks	Over 280 Brinell after heating and water quench
	414	Martensitic		Long white with few forks	
	416	Martensitic		Long white with few forks	
	416Se	Martensitic			
	420	Martensitic		Long white-red with burst	
	431	Martensitic		Long white with few forks	
	440, A, B, C	Martensitic		Long white-red with burst	
	430	Ferritic		Long white with few forks	180-250 Brinell after heating to 1800"F and water quench
	430F	Ferritic			
	430FSe	Ferritic			
	446	Ferritic		Full red without many forks	

(Continued on next page)

Table 5-5 continued

	AISI TYPE	GROUP	SULFURIC ACID TEST	HYDROCHLORIC ACID TEST
CHROMIUM-NICKEL GRADES	302	Austenitic	Strong attack, Dark surface, Green crystals	Fairly rapid reaction Pale blue-green solution
	303	Austenitic		Spoiled egg odor, heavy black smudge
	303Se	Austenitic		Garlic odor
	304	Austenitic	Strong attack, Dark surface, Green crystals	Fast attack Gas formation
	308	Austenitic		
	309	Austenitic		
	310	Austenitic		
	316	Austenitic	Slow attack, Tan surface turns brown	Very slow attack compared to 302, 304, 321 and 347
	317	Austenitic	Slower attack, Tan surface turns brown	
	321	Austenitic		Fast attack Gas formation
	347	Austenitic		Fast attack Gas formation
CHROMIUM GRAD	410	Martensitic		More vigorous reaction than 302 Darker green solution
	414	Martensitic		
	416	Martensitic		Spoiled egg odor
	416Se	Martensitic		Garlic odor
	420	Martensitic		
	431	Martensitic		
	440, A, B, C	Martensitic		
	430	Ferritic		
	430F	Ferritic		Spoiled egg odor
	430FSe	Ferritic		Garlic odor
	446	Ferritic		

6

MISCELLANEOUS ITEMS

There are small items such as screwed fittings, gaskets, pipe nipples, and plugs that are necessary on every project. The materials described in this chapter are merely to acquaint the new materials person of their existence, and so are not shown with dimensions and tables.

Malleable iron fittings, either black or galvanized, are the low-pressure fittings sold at the hardware store, and used accordingly. Figure 6-1 shows a 150-lb bronze to iron ground joint union. This type fitting is available from 1/8-in. through 4-in. in 150, 250, and 300-lb ratings.

Figure 6-1. Malleable iron union. (Courtesy of Jaqua-McKee, Inc.)

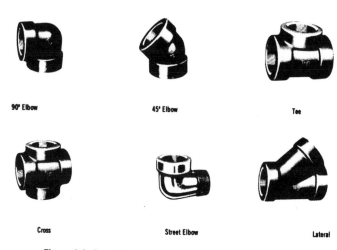

90° Elbow   45° Elbow   Tee

Cross   Street Elbow   Lateral

Figure 6-2. Forged steel fittings. (Courtesy of Jaqua-McKee, Inc.)

Figure 6-2 shows the most commonly used fittings in ANSI systems, the forged steel screwed (and socket-weld) fittings. These fittings are used for steam, water, oil, gas, and air. They are available in 2,000, 3,000 and 6,000-lb classes, in many types of alloys including stainless steels. Figure 6-3 shows additional forged steel fittings. Store fittings by size and rating.

Pipe nipples are stocked in various lengths in black or galvanized pipe, and in schedules to match the pipe being used. They are also made to the length required for fit-up by pipefitters using pipe machines. Figure 6-4 depicts typical pipe nipples. One of the nipples is referred

Coupling **Half Coupling** **Reducer** **Cap**

Hex. Head Bushing **Flush Bushing** **Round Head Plug** **Square Head Plug**

Hex. Head Plug

Figure 6-3. Forged steel fittings. (Courtesy of Jaqua-McKee, Inc.)

All Thread Nipple **Nipple Threaded Both Ends**

Figure 6-4. Pipe nipples. (Courtesy of Jaqua-McKee, Inc.)

to as an all-thread or close nipple by the craftsmen. Other nipples are described in size by the nipple's length. A *shoulder nipple* is an all-thread nipple with a small unthreaded section in the center of the nipple forming a shoulder. Some nipples have a thread on one end only. Some used with socketweld fittings do not have any threads. Store nipples by type, size, and length.

Swage nipples are used to reduce pipe sizes. Swages are available in combinations from 1/8-in. to about 8-in., but larger in special cases. Figure 6-5 shows a typical swage nipple threaded on both ends. Swages may have almost any combination of ends such as bevel large end, thread small end, grooved, and bevel both ends.

Another type of cross-over fitting is the *sub-tubing nipple* as shown in Figure 6-6. Sub-tubing nipples are used to change from an API tubing thread to a line pipe thread. The longer variations of this nipple are called

Figure 6-5. Swage Nipples. (Courtesy of Jaqua-McKee, Inc.)

Figure 6-6. Sub-tubing nipple, pup-joints. (Courtesy of Jaqua-McKee, Inc.)

Figure 6-7. Bull plugs. (Courtesy of Jaqua-McKee, Inc.)

pup-joints. Pup-joints are in even-numbered lengths from 2-ft through 16-ft, and are used to complete a string of pipe in given length without cutting and threading. Pup-joints are usually API threads on both ends.

Bull plugs, see Figure 6-7, are used to close ends of lines or strings of tubing in oil wells. They are akin to the smaller hex head or round pipe plugs, but are not the same. Bull plugs are available in sizes from 1/8-in. through 8-in., and can have threaded, plain, beveled, or grooved ends. Bull plugs are often drilled and tapped, and a valve and nipple added for an outlet. There are female threaded plugs available, but they are not common.

131

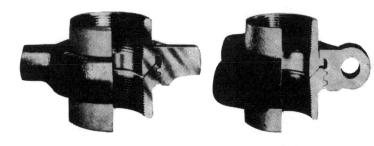

Figure 6-8. Hammer unions. (Courtesy of Jaqua-McKee, Inc.)

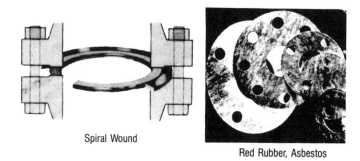

Spiral Wound

Red Rubber, Asbestos

Figure 6-9. Typical ANSI gaskets. (Courtesy of Jaqua-McKee, Inc.)

Figure 6-8 shows the *wing* or *hammer union.* These rugged unions are used most commonly in oil field hook-ups and temporary lines. They are available in sizes from 1-in. through 12-in., from 1,000 psi to 15,000 psi.

132

Gaskets require proper storage methods for protection. Do not store gaskets on nails or pegs. Store by type, size, and rating on a flat surface. Never issue gaskets that will be placed in a sack with the fittings and bolts. Figure 6-9 illustrates some of the common ANSI system gaskets.

Flange insulation sets are used between systems, such as an offshore pipeline connection to the production facilities. The set consists of a flange gasket, either full faced or raised face, sleeves, and washers for the bolts. Store the insulation kits in the same vendor box as they were shipped in. Do not mix or mingle the various sets. Figure 6-10 is a flange insulation kit or set.

Corrosive areas sometimes mandate the use of a protector of the flange stand-off area to protect the gasket area between the flanges. These protectors are available

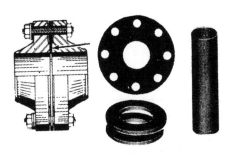

Figure 6-10. Flange insulation sets. (Courtesy of Jaqua-McKee, Inc.)

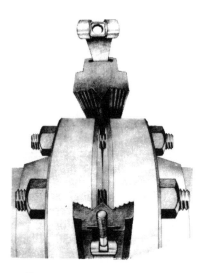

Figure 6-11. Process flange protectors. (Courtesy of Rodun Development Corp., Houston, Texas.)

in various shapes and models. Figure 6-11 shows a protector called Flexi-seal.® (Rodun Development Corp., Houston, Texas). This protector features a center V-shaped section that forms its primary seal. Flexible ribs on each side provide secondary sealing. It is secured in place by a stainless steel band and latched with a steel T-bolt.

Store miscellaneous items by size and rating, and if necessary, apply a tag for future identification.

7

PIPELINE PIGS

Pipelines require cleaning, and products require separation when being transported through the same line simultaneously; pipeline pigs are used to make cleaning and separation possible.

The slang word "pig" means scraper, ball, sphere, or other apparati used in pipelines. Figure 7-1 illustrates a Polly-Cast® pig made from polyester urethane. Pigs such as the Polly-Cast® can be used in gas, crude oil, salt water, refined product, and LPG lines. They are also available with steel bristles on the wear surface, which increases the life of the urethane pig.

Spherical balls or pigs (Figure 7-2) have been used for many years. They are available in solid material in sizes 1″ through 12″ and inflatable from 4″ through 56″. Pigs are inflated with a displacement pump capable of 250 psi or 1,724 kPa (Figure 7-3). This type of pig is often used in automated piping systems, meter proving, product separation, hydrostatic testing, and cleaning. Most are made from Neoprene, Nitrile, Polyurethane, or Viton.

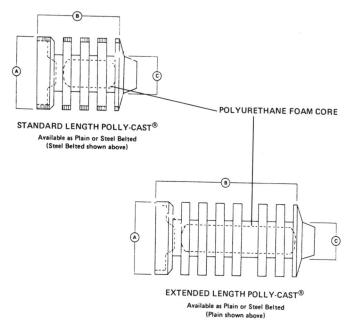

Figure 7-1. Polly-Cast pig. (Courtesy of Knapp Polly Pig, Inc.)

Without pigging, product buildup, sediment, and slime in pipelines increases and flow capacity decreases. Eventually, the pipeline might even need to be replaced entirely. The cost of the power needed to move the products also increases dramatically as buildup increases. Without pigs for product separation, separate pipelines would be required for different fluids. Pigs are a routine part of pipeline operations.

136

Figure 7-2. Spherical pipeline pigs. (Courtesy of LTV Energy Products Company.)

Figure 7-3. Pressure pump. (Courtesy of LTV Energy Products Company.)

137

Pigs are placed into lines by means of a launcher and a receiver, as shown in Figures 7-4a and b. They are moved down the line by means of pressure. Both the launcher and the receiver have bolted, swinging doors to install or remove the pigs. In the case of very large diameter pigs, a rail hoist is part of both units. Pigs are usually ⅛" larger than the inside diameter of the pipeline for a good tight fit. At intersections of pipelines, Scraper Bar Tees (Figure 2-15) are used to keep the pig in line. The pig moves right through bends and turns until it reaches the receiver. Near the end of its trip and at checkpoints on the way, the pig passes over a device known as a "pig signal," and raises

Figure 7-4A. Pipeline pig launcher. (Courtesy of Tube Turns Technologies, Inc.)

138

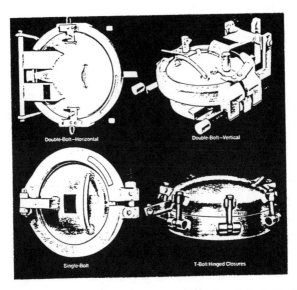

Figure 7-4B. Pipeline pig closures. (Courtesy of Tube Turns Technologies, Inc.)

either a flag notice on a manual system or an indicating light on a panel board. (See Figure 7-5). Pigs are tracked in lines by a radioactive isotope which gives off a radio signal and can be monitored from the air, ground, or a satellite tracked by an electronic transmitter.

The most common pig is the polyurethane style used for pigging lines up to 60 inches. These pigs are used for product or crude oil pipelines, chemical process piping, water systems, offshore condensate removal, and many other variations of pigging.

Figure 7-5. Automated closures. (Courtesy of Tube Turns Technologies, Inc.)

A noninflatable Polly-Sphere pig (Figure 7-6) has a high-density polyurethane foam core and a hard but flexible outer cover. This type of sphere offers the advantage of not deflating and the pressures inside the sphere and in the line are equalized by small holes drilled in the pig. It can be used for low pressure lines as well.

Bullet-shaped pigs are made of durable foam. A special exterior surface made from plastic, in a spiral or

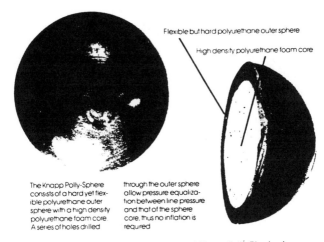

Flexible but hard polyurethane outer sphere

High density polyurethane foam core

The Knapp Polly-Sphere consists of a hard yet flexible polyurethane outer sphere with a high density polyurethane foam core. A series of holes drilled through the outer sphere allow pressure equalization between line pressure and that of the sphere core, thus no inflation is required

Figure 7-6. Polly-Sphere pig. (Courtesy of Knapp Polly Pig, Inc.)

criss-cross design imparts greater cleaning power, strength, and wearability. The nose is completely covered to insure proper sealing. For extra-tough cleaning jobs, a coat of silicon carbide or abrasives can be added. Flame-hardened steel wires are bonded to the polyurethane body of the pig (Figure 7-7) for use on long runs. The bristles are mounted at the specific angle that makes them self-sharpening. A simpler pig, the foam Sweege Pig, is made of extremely soft foam and is designed for temporary sweeping of lines to eliminate things such as oxides which cause red water. They are propelled by normal water pressure. A Polly-Pig of this type is shown in Figure 7-8.

141

Havelina Polly Pig

Super Havelina Polly Pig

Coated Super
Havelina Polly Pig

Figure 7-7. Bullet-shaped pigs. (Courtesy of Knapp Polly Pig, Inc.)

Inexpensive Foam Sweege Pigs are now available to temporarily remove oxides that cause red water.

The standard Sweege Foam Pig, Style $\underline{V}$-B, is an inexpensive, expendable Knapp Polly-Pig® of extremely soft foam designed specifically for temporary sweeping of a line to eliminate oxides that cause red water. They are easily inserted into water mains by hand, require no special devices, no line shut offs, and are propelled by normal water pressure.

Figure 7-8. Foam sweege pig. (Courtesy of Knapp Polly Pig, Inc.)

The diameter of Polly-Pigs is usually one and a half times the length. As a rule, ⅛″ is added to the pig body size for coating. Some pigs with diameters less than 6″ have mesh for internal reinforcement. Some are equipped with a rope on one or both ends for ease in handling and pulling. Dished ends add to the effective fluid removal while pointed ends are used for bi-directional changes. The ends may be either shorter or longer for extreme valves or launches. They also come in either soft or hard for variations in density (Figure 7-9), which is measured in pounds per cubic foot.

Some styles of Polly-Pigs made by Knapp are described in Figure 7-10. They are color-coded by type of cover for identification and some are criss-crossed.

Single spiral coating pattern allows greater flexibility of the pig for running tight bends. "T s." valves and variations in pipe I D

Double spiral coating pattern provides greater number of cleaning edges and makes the pig more resistant to tearing. This design is intended for longer runs where less flexibility is required

Silicon carbide impregnated coating is available in both single and double spiral patterns. This coating is advisable when removing mill scale, weld slag, calcite or carbonate deposits, rust, or other hard or abrasive deposits.

Figure 7-9. Coated polly pigs. (Courtesy of Knapp Polly Pig, Inc.)

Urethane Scraper Cup specifications from 3″ through 56″ are described in Table 7-1. It is important to note that cups up to 14″ are molded without a center hole, so you must specify the hole size required when ordering. The approximate pressures and flows required for polly pigging are shown in Table 7-2.

(Text continued on page 148)

STYLE	TYPE	DENSITY	FUNCTION
	SBD (Scarlet bare durafoam)	8 LBS./CU. FT.	Heavy Drying Up to 200 Ml
	SCC (Scarlet criss-cross)	8 LBS./CU. FT.	Heavy Wiping Up to 200 Ml
	SCC-WB (Scarlet criss-cross wire brush)	8 LBS./CU. FT.	Heavy Scraping Up to 200 Ml
	SCC-SC (Scarlet criss-cross silicon carbide)	8 LBS./CU. FT.	Heavy Scraping Up To 200 Ml
	SBD-T (Turning)	8 LBS./CU. FT.	Heaviest Drying Up To 300 Ml
	SCC-T (Turning)	8 LBS./CU. FT.	Heaviest Wiping Up To 300 Ml
	SCC-WB-T (Turning)	8 LBS./CU. FT.	Heaviest Scraping Up To 300 Ml
	SCC-SC-T (Turning)	8 LBS./CU. FT.	Heaviest Scraping Up To 300 Ml
	RBS (Red bare squeegee)	5 LBS./CU. FT.	Regular Drying Up To 10 Ml
	RCC (Red criss-cross)	5 LBS./CU. FT.	Regular Wiping Up To 10 Ml
	RCC-WB (Red criss-cross wire brush)	5 LBS./CU. FT.	Regular Scraping Up To 10 Ml

Figure 7-10. Polly pig styles. (Courtesy of Knapp Polly Pig, Inc.)

STYLE	TYPE	DENSITY	FUNCTION
	RCC-SC (Red criss-cross silicon carbide)	5 LBS./CU. FT.	Regular Scraping Up To 10 MI
	RBS-T (Turning)	5 LBS./CU. FT.	Longer Drying Up To 25 MI
	RCC-T (Turning)	5 LBS./CU. FT.	Longer Wiping Up To 25 MI
	RCC-WB-T (Turning)	5 LBS./CU. FT.	Longer Scraping Up To 25 MI
	RCC-SC-T (Turning)	5 LBS./CU. FT.	Longer Scraping Up To 25 MI
	YBS (Yellow bare swab)	2 LBS./CU. FT.	Light Drying Up To 1 MI
	YCC (Yellow criss-cross)	2 LBS./CU. FT.	Light Drying Up To 1 MI
	YCC-SC (Yellow criss-cross silicon carbide)	2 LBS./CU. FT.	Light Drying Up To 1 MI
	YBS-B (Bullet)	2 LBS./CU. FT.	Light Drying Up To 1 MI
	YCC-T (Turning)	2 LBS./CU. FT.	Light Drying Up To 3 MI
	YCC-SC-T (Turning)	2 LBS./CU. FT.	Light Drying Up To 3 MI
	UNICAST	20 LBS./CU. FT.	Long Range cleaning Up To 2000 MI
	GRAY HARD SCALE	8 LBS./CU. FT.	Industrial Scraping Up To 300 MI
	MAXI-BRUSH (LIGHT WIRE)	8 LBS./CU. FT.	Maximum Scraping Up To 300 MI
	MAXI-BRUSH (HEAVY WIRE)	8 LBS./CU. FT.	Maximum Scraping Up To 300 MI

Table 7-1
Urethane Scraper Cups

	STANDARD & STEEL BELTED POLLY-CUP®									
NOM. PIPE SIZE	A Size in Inches	B Size in Inches	C Size in Inches	D Size in Inches	E Size in Inches	F Size in Inches	G Size in Inches	HØ Size in Inches	K # HOLES	APPX. WT. LBS.
3"	3.19	*	1.13	*	.50	2.25	2.75	*	*	.3
4"	4.25	*	1.38	*	.50	3.07	3.38	*	*	.4
6"	6.13	*	1.75	*	.50	4.25	5.25	*	*	1.3
8"	8.25	*	2.00	*	.75	6.13	7.25	*	*	2.6
10"	10.50	*	2.25	*	.88	8.13	8.75	*	*	4.8
12"	12.22	*	2.25	1.63	.88	12.07	11.50	*	*	7.3
14"	14.25	*	2.50	1.88	.88	13.00	11.63	*	*	8.7
16"	16.00	8.75	2.75	2.25	1.00	15.25	13.88	11.00	8	9.9
18"	18.00	10.88	3.00	2.38	1.00	17.25	16.00	12.75	8	12.2
20"	20.25	10.88	3.50	2.63	1.25	19.00	17.25	13.50	12	18.6
22"	†	†	†	†	†	†	†	†	†	24.4
24"	24.25	14.13	4.00	2.88	1.50	23.25	21.63	16.50	14	29.8
26"	†	†	†	†	†	†	†	†	†	34.8
28"	†	†	†	†	†	†	†	†	†	
30"	30.25	16.13	4.38	3.00	1.88	29.50	26.25	19.50	16	51.0
32"	†	†	†	†	†	†	†	†	†	57.5
34"	†	†	†	†	†	†	†	†	†	74.0
36"	36.25	22.13	4.38	3.00	1.88	35.88	33.25	25.00	22	77.0
40"	40.25	22.13	5.00	3.25	2.00	38.50	35.75	26.00	24	80.8
42"	42.25	24.13	5.50	3.38	2.00	†	†	†	24	100.0
48"	48.25	32.13	6.00	3.50	2.00	46.00	44.50	36.00	24	105.0
56"	56.25									144.0

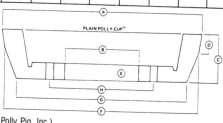

PLAIN POLLY-CUP®

(Courtesy Knapp Polly Pig, Inc.)

Table 7-2
Pressures Required for Polly Pigging
APPROXIMATE PRESSURES AND FLOWS REQUIRED FOR POLLY PIGGING

NOMINAL PIPE I.D.		PSI	BARS	GPM	LPM	CFM	M³/M
2"	5.0cm	100-200	7.0-14.0	100-200	380-700	15-25	.4-.7
3"	7.6cm	100-150	7.0-10.5	150-300	600-1150	75-100	2.1-2.8
4"	10.0cm	75-125	5.2-8.8	250-400	950-1500	125-175	3.5-5.0
6"	15.2cm	50-100	3.5-7.0	450-600	1700-2275	200-250	5.6-7.0
8"	20.0cm	30-80	2.1-5.6	650-800	2500-3050	275-350	7.7-10.0
10"	25.4cm	20-60	1.4-4.2	700-1000	2650-3800	375-450	10.5-13.0
12"	30.4cm	10-50	0.7-3.5	800-1200	3050-4550	450-550	13.0-15.5
14"	35.4cm	10-40	0.7-2.8	1000-1400	3800-5300	550-650	15.5-18.3
16"	40.6cm	5-35	0.35-2.4	1200-1600	4550-6050	625-750	17.5-21.1
18"	45.4cm	5-30	0.35-2.1	1600-1800	6050-6825	725-850	20.3-23.9
20"	50.5cm	5-25	0.35-1.7	1500-2200	5700-7575	825-950	24.5-26.7
22"	55.0cm	5-25	0.35-1.7	1700-2200	5450-8550	900-1050	35.2-29.5
24"	60.0cm	5-20	0.35-1.4	1900-2400	7200-9000	1050-1200	29.4-32.3
26"	65.2cm	5-20	0.35-1.4	2100-2600	7950-9750	1150-1300	32.2-35.1
28"	70.0cm	5-20	0.35-1.4	2300-2800	8700-10600	1250-1400	35.0-39.5
30"	76.2cm	5-10	0.35-0.7	2400-3000	10000-11350	1300-1500	36.4-42.3
36"	91.4cm	5-10	0.35-0.7	3000-3600	11350-13650	1800-2000	50.4-56.6
40"	101.1cm	5-10	0.35-0.7	3400-4000	12900-15150		

(Courtesy Knapp Polly Pig, Inc.)

(Text continued from page 144)

Pigs should be stored indoors where possible, on their ends and sorted by size and style. Staples, pins, or other sharp objects should not be used to attach tags or bar codes. An easily removable self-adhesive bar code tag is recommended instead.

PIPING ABBREVIATIONS

API	American Petroleum Institute	AWWA	American Water Works Association
ASA	American Standards Association	BE	Beveled end
		BW	Buttweld
Asb	Asbestos (gaskets)	BBE	Bevel both ends
ANSI	American National Standards Institute, Inc.	Bbl	Barrel
		Bdr	Bleed ring
		Bfy	Butterfly (valve)
		Bld	Blind (flange)
ASME	The American Society of Mechanical Engineers	BLE	Bevel large end
		Blk	Black (pipe)
		BOE	Bevel one end
ASTM	The American Society for Testing and Materials	BOM	Bill of materials
		BOP	Bottom of pipe

Abbreviation	Meaning	Abbreviation	Meaning
Brz	Bronze (valve)	Ditto	Do not use this term.
BSE	Bevel small end	DSAW	Double sub-merged Arc Welded (pipe)
CI	Cast iron		
Cm	Centimeter		
CS	Cast Steel, carbon steel, cap screw	Dwg #	Drawing Number
		Ea.	Each
Cu	Cubic	El	Elevation (on drawing)
CW	Chain wheel		
Chk	Check (valve)	Ecc	Eccentric
Cpl	Coupling	Ell	Elbow
CSC	Car seal closed	Eol	Elbolet®
		ERW	Electric Resistance Weld (pipe)
Csg	Casing		
CSO	Car seal open	Esw	Eccentric Swage
Csw	Concentric swage		
CWO	Chain wheel operator	EUE	External upset ends
CWP	Cold water pressure	Ex. hvy	Extra heavy
		Ex. stg	Extra strong
Conc	Concentric	Exp jt	Expansion joint
DI	Ductile iron		
D&T	Drill & Tap	Elec	Electrical
D&W	Doped & Wrapped (pipe)	F&D	Faced and drilled (flange)
		FE	Flanged ends/ Flow element
DES	Double extra strong	FF	Flat/Full face
Dia.	Diameter	F/F	Face of flange
Dim.	Dimension	FS	Forged steel
		Ft	Feet/Foot

150

FW	Field weld/Firewater	ISO	Isometric (drawing)
FAB	Fabricate/Fabricator	IUE	Internal upset ends
FAS	Free along side	IS&Y	Inside screw & yoke (valve)
Fem	Female (ends)	IBBM	Iron body bronze mounted (valve)
Fig	Figure (number)		
Flg.	Flange	Insl	Insulation
FOB	Free on board	Jt (s)	Jt. (Joints)
FSD	Flat side down	JW	Jacket water
FSU	Flat side up	Jkscr	Jack screw
Flex	Flexitallic (gasket brand name)	Lb (s), #	Pound (s) #symbol for pounds
Flgd	Flanged	Lg	Length, long, level gauge
GG	Gauge glass		
GJ	Ground joint (union)	LJ	Lap joint (flange)
Gal	Gallon	LP	Line pipe
Glb	Globe (valve)	LR	Long radius
Gsk	Gasket	LLC	Liquid level controller
Galv	Galvanized		
HN	Heat number	LOL	Latrolet®
Hdr	Header	Latl	Lateral
Hex	Six-sided head, bolt, plug, etc.	Lin Ft	Linear feet
		M	Meter/one thousand
Hvy	Heavy	MI	Malleable iron
ID	Inside diameter	Mk	Mark (spool piece)
IPS	Iron pipe size		

MM	Millimeter	OD	Outside diameter
M&F	Male & Female (ends)	Oz	Ounce
Max	Maximum (a warehouse stocking level)	Orf	Orifice
		OS&D	Over short & damage (report)
Mfg	Manufacturer		
Min	Minimum (a warehouse stocking level)	OS&Y	Outside screw & yoke (valve)
Misc	Miscellaneous (schedules of pipe)	Pc	Piece (mark for spool pieces)
		PE	Plain ends
MRR	Materials receiving report	PI	Pressure indicator (valves & gauge assembly)
MSS	Manufacturers Standards Society of the Valve and Fittings Industry		
		PO #	Purchase order or number
MTO	Material Take-off (from drawings)	Pr	Pair of items
		PS	Pipe support
		PW	Potable water
NC	Normally closed	PBE	Plain both ends
No or #	Number	Pdl	Paddle (a blind plate between flanges)
NU	Non-upset (ends)		
Nip	Nipple (pipe)	PLE	Plain large end
NPS	Nominal pipe size	Plt	Plate (steel)
		POE	Plain one end
NPT	Nominal pipe thread	PSE	Plain small end

psi	Pounds per square inch	SE	Screwed ends
PVF	Pipe, valves and fittings	SO	Slip-on (flange)
PSV	Pressure safety (relief) valve	Sq	Square feet, yards, etc.
Press	Pressure	SR	Short radius, stress relieve
psig	Pounds-force square inch, gauge	SS	Stainless steel
		SW	Socket weld
Qty	Quantity	S/40	Schedule 40 (of pipe or fittings)
RF	Raised face		
RR	Red rubber (gasket type)	Sch	Schedule (of pipe or fittings)
RS	Rising stem (valve)	Sdl	Saddle (pipe)
Rad	Radius	SOL	Sockolet®
Red	Reducer	SRL	Short radius ell
RPM	Revolutions per minute	Std	Standard (a pipe or fitting schedule)
RTE	Reducing tee		
Rtg	Rating	Stl	Steel
RTJ	Ring type joint (flange facing)	Stm	Steam
		Sub	Short length of pipe or rod
Rdcr	Reducer		
Rec'd	Received (materials)	Swg	Swage nipple
Req'd	Required (materials)	SWP	Safe working pressure
Reqn	Requisition	Skt Bld	Skillet blind (plate between flanges)
SC	Sample connection		

Spl Sht	Spool sheet (from isometric drawing)	TOL	Thredolet®
		TSE	Thread small end
Scrd	Screwed (ends)	Typ	Typical (repeat the same item)
Smls	Seamless		
Spec		Thr'd	Threaded
bld	Spectacle blind (plate between flanges)	Un	Union
		Va	Valve
		Vac	Vacuum
		Vol	Volume
TI	Temperature indicator	Wd	Width/Wide
		WE	Weld end
TW	Thermometer well	WI	Wrought iron
		WN	Weldneck (flange)
T&C	Threaded & coupled	WP	Working pressure
T&G	Tongue & groove (flange facing)	WT	Wall thickness/weight
Tbg	Tubing	WOL	Weldolet®
Thk	Thick	XR	X-ray (at pipe welds)
TLE	Thread large end	XS	Extra strong
		XXS	Double extra strong
TOE	Thread one end	Yd	Yard

APPENDIX B

USEFUL FORMULAS

For freight: Cubic Feet = Height (in.) × Width (in.) × Length (in.) ÷ 1,728

      2000 pounds = Short ton
      2240 pounds = Long ton
      40 cubic feet = Measurement ton
      Metric ton = 1.1 tons

For concrete: Length (ft) × Width (ft) × Height (ft) ÷ 27 = Cubic yards

Miscellaneous Factors:

12 inches	=	1-foot
2.54 centimeters	=	1 inch
144 inches	=	1 square foot
10.764 square feet	=	1 square meter
3 feet	=	1 yard
9 square feet	=	1 square yard
3.2808 feet	=	1 meter
1728 inches	=	1 cubic foot

```
    27 cubic feet = 1 cubic yard
0.62137 miles = 1 kilometer
        1 mile = 63,360 inches, 5,280 feet, 1760
                 yards
        1 acre = 43,560 square feet
    16 ounces = 1 pound
0.45359 kilograms = 1 pound
```

Converting English and Metric Units.

Reading and Converting English and Metric Units

English: (Inches are equally divided into 16 parts of $1/16''$)

$1/16''$	$= 0.625''$				
$1/8''$	$= 2/16''$	$= 0.125''$			
$1/4''$	$= 2/8''$	$= 0.25''$			
$1/2''$	$= 2/4''$	$= 4/8''$	$= 8/16''$	$= 0.5''$	
$1''$	$= 2/2''$	$= 4/4''$	$= 8/8''$	$= 1.0''$	

Example:

$$6'' + 1/4'' + 1/16'' =$$
$$6'' + 4/16'' + 1/16'' = 6\,5/16''$$

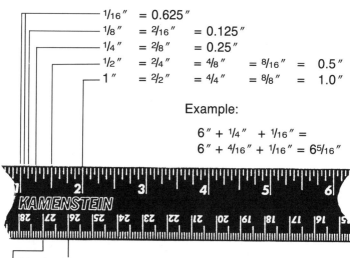

Metric: (Centimeters are equally divided into ten parts of 1 millimeter)

```
1cm = 10mm =  .01m
1mm = 0.1cm = 0.001m
```

Example:

—14cm + 4mm = 14.4cm = 144m

English to Metric Conversions:

```
1'     = 30.48cm
1"     = 2.54cm
1/2"   = 12.7mm
1/4"   = 6.35mm
1/8"   = 3.175mm
1/16"  = 1.588mm
```

Metric to English Conversions:

```
1cm    = 0.39"
0.5cm  = 0.195"
1mm    = 0.039"
```

Courtesy of Abby Dawkins, West Hartford, CT.

INDEX

160